WORKSH

DEANA RICHMOND

BEGINNING ALGEBRA WITH APPLICATIONS AND VISUALIZATION

THIRD EDITION

Gary Rockswold

Minnesota State University, Mankato

Terry Krieger

Rochester Community and Technical College

PEARSON

oston Columbus Indianapolis New York San Francisco Upper Saddle River
Amsteam Cape Town Dubai London Madrid Milan Munich Paris Montreal Toronto
Del Mexico City Sao Paulo Sydney Hong Kong Seoul Singapore Taipei Tokyo

Reproduced by Pearson from electronic files supplied by the author.

Publishing as Pearson, 75 Arlington Street, Boston, MA 02116.

ISBN-13: 978-0-321-74802-7
ISBN-10: 0-321-74802-6

www.pearsonhighered.com

PEARSON

Contents

Name: ____________________ **Course/Section:** ____________________ **Instructor:** ____________________

Chapter 1 Introduction to Algebra
1.1 Numbers, Variables, and Expressions

Natural Numbers and Whole Numbers ~ Prime Numbers and Composite Numbers ~ Variables, Algebraic Expressions, and Equations ~ Translating Words to Expressions

STUDY PLAN

Read: Read Section 1.1 on pages 2-8 in your textbook or eText.

Practice: Do your assigned exercises in your ☐ Book ☐ MyMathLab ☐ Worksheets

Review: Keep your corrected assignments in an organized notebook and use them to review for the test.

Key Terms

Exercises 1-10: Use the vocabulary terms listed below to complete each statement. Note that some terms or expressions may not be used.

product	**algebraic expression**
equation	**whole number**
formula	**composite number**
variable	**natural number**
factor	**prime number**
	prime factorization

1. A(n) ____________________ is a special type of equation that expresses a relationship between two or more quantities.

2. The set of ____________________(s) may be expressed as $0, 1, 2, 3, 4, 5, 6, \ldots$.

3. A natural number greater than 1 that has only itself and 1 as natural number factors is a(n) ____________________.

4. Multiplying natural numbers 3 and 4 results in 12, a natural number. The result 12 is called the ____________________ and the numbers 3 and 4 are ____________________(s) of 12.

5. The expression $2 \cdot 2 \cdot 2 \cdot 3 \cdot 5$ is the ____________________ of 120.

6. A(n) ________________ is a symbol, typically an italic letter such as x, y, z or F, used to represent an unknown quantity.

7. The set of ________________(s) comprise the counting numbers and may be expressed as 1, 2, 3, 4, 5, 6, . . .

8. A(n) ________________ consists of numbers, variables, operation symbols such as $+, -, \cdot,$ and $\div$, and grouping symbols such as parentheses.

9. A natural number greater than 1 that has factors other than itself and 1 is a(n) ________________.

10. A(n) ________________ is a mathematical statement that two algebraic expressions are equal.

Prime Numbers and Composite Numbers

Exercises 1-4: Refer to Example 1 on page 4 in your text and the Section 1.1 lecture video.

Classify each number as prime or composite, if possible. If a number is composite, write it as a product of prime numbers.

1. 29 **1.** ________________

2. 0 **2.** ________________

3. 65 **3.** ________________

4. 180 **4.** ________________

Variables, Algebraic Expressions, and Equations

Exercises 5-14: Refer to Examples 2-4 on pages 5-6 in your text and the Section 1.1 lecture video.

Evaluate each algebraic expression for the given value of x.

5. $x+6;\ x=4$ **5.** ________________

6. $4x;\ x=3$ **6.** ________________

7. $17-x;\ x=2$ **7.** ________________

8. $\frac{x}{(x-4)};\ x=8$ **8.** ________________

Evaluate each algebraic expression for the given values of y and z.

9. $4yz;\ y=1,\ z=5$ **9.** ________________

10. $z-y;\ y=7,\ z=1$ **10.** ________________

11. $\frac{z}{y};\ y=6,\ z=18$ **11.** ________________

Find the value of y for the given value(s).

12. $y=x+2;\ x=13$ **12.** ________________

13. $y=\frac{x}{5};\ x=5$ **13.** ________________

14. $y=4xz;\ x=2,\ z=3$ **14.** ________________

Translating Words to Expressions

Exercises 15-20: Refer to Examples 5-7 on pages 6-7 in your text and the Section 1.1 lecture video.

Translate each phrase to an algebraic expression. Specify what each variable represents.

15. Five less than a number **15.** ______________

16. Three times the cost of a DVD **16.** ______________

17. A number minus 5, all multiplied by a different number **17.** ______________

18. The quotient of 100 and a number **18.** ______________

19. For each year after 2000, the average price of a gallon of milk has increased by about $0.15 per year.

(a) What was the total increase in the price of a gallon of milk after 10 years, or in 2010? **19. (a)**______________

(b) Write a formula (or equation) that gives the total increase P in the average price of a gallon of milk t years after 2000. **(b)**______________

(c) Use your formula to estimate the total increase in the price of a gallon of milk in 2020. **(c)**______________

20. The volume V of a rectangular box equals its length l times its width w times its height h.

(a) Write a formula that shows the relationship between these four quantities. **20. (a)**______________

(b) Find the volume of a rectangular box that is 12 inches long, 5 inches wide, and 3 inches high. **(b)**______________

Name: **Course/Section:** **Instructor:**

Chapter 1 Introduction to Algebra
1.2 Fractions

Basic Concepts ~ Simplifying Fractions to Lowest Terms ~ Multiplication and Division of Fractions ~ Addition and Subtraction of Fractions ~ An Application

STUDY PLAN

Read: Read Section 1.2 on pages 11-23 in your textbook or eText.

Practice: Do your assigned exercises in your ☐ Book ☐ MyMathLab ☐ Worksheets

Review: Keep your corrected assignments in an organized notebook and use them to review for the test.

Key Terms

Exercises 1-5: Use the vocabulary terms listed below to complete each statement. Note that some terms or expressions may not be used.

reciprocal
greatest common factor (GCF)
basic principle of fractions
multiplicative inverse
lowest terms
least common denominator (LCD)

1. A fraction is said to be in ____________________ when the numerator and denominator have no factors in common.

2. The ____________________ for two or more fractions is the smallest number that is divisible by every denominator.

3. The ____________________ states that the value of a fraction is unchanged if the numerator and denominator of the fraction are multiplied (or divided) by the same nonzero number.

4. The ____________________ or ____________________ of a nonzero number a is $\frac{1}{a}$.

5. The ____________________ of two or more numbers is the largest factor that is common to those numbers.

Basic Concepts

Exercises 1-3: Refer to Example 1 on page 12 in your text and the Section 1.2 lecture video.

Give the numerator and denominator of each fraction.

1. $\frac{7}{15}$ **1.** ____________

____

2. $\frac{x}{yz}$ **2.** ____________

____

3. $\frac{a+2}{b-3}$ **3.** ____________

____

Simplifying Fractions to Lowest Terms

Exercises 4-7: Refer to Examples 2-3 on pages 13-14 in your text and the Section 1.2 lecture video.

Find the greatest common factor (GCF) for each pair of numbers.

4. 28, 70 **4.** ____________

5. 32, 40 **5.** ____________

Simplify each fraction to lowest terms.

6. $\frac{18}{45}$ **6.** ____________

7. $\frac{45}{105}$ **7.** ____________

Multiplication and Division of Fractions

Exercises 8-20: Refer to Examples 4-8 on pages 15-18 in your text and the Section 1.2 lecture video.

Multiply. Simplify the result when appropriate.

8. $\frac{3}{5} \cdot \frac{4}{7}$ **8.** ________________

9. $\frac{6}{7} \cdot \frac{2}{3}$ **9.** ________________

10. $4 \cdot \frac{5}{8}$ **10.** ________________

11. $\frac{a}{b} \cdot \frac{c}{5}$ **11.** ________________

Find each fractional part.

12. Three-fourths of one-third **12.** ________________

13. Two-thirds of one-fifth **13.** ________________

14. Three-fourths of eight **14.** ________________

15. Approximately nine-tenths of Team in Training participants are female. About one-third of those are over the age of 40. What fraction of Team in Training participants are women over the age of 40? **15.** ________________

Divide. Simplify the result when appropriate.

16. $\frac{1}{4} \div \frac{4}{7}$ **16.** ______________

17. $\frac{2}{3} \div \frac{2}{3}$ **17.** ______________

18. $6 \div \frac{12}{5}$ **18.** ______________

19. $\frac{x}{3} \div \frac{z}{y}$ **19.** ______________

20. Describe a problem for which the solution could be found by dividing 3 by $\frac{1}{4}$. **20.** ______________

Addition and Subtraction of Fractions

Exercises 21-29: Refer to Examples 9-12 on pages 19-22 in your text and the Section 1.2 lecture video.

Add or subtract as indicated. Simplify your answer to lowest terms when appropriate.

21. $\frac{5}{11} + \frac{7}{11}$ **21.** ______________

22. $\frac{13}{9} - \frac{7}{9}$ **22.** ______________

Find the LCD for each set of fractions.

23. $\frac{2}{3}, \frac{1}{4}$ **23.** ____________

24. $\frac{4}{15}, \frac{3}{20}$ **24.** ____________

Rewrite each set of fractions using the LCD.

25. $\frac{2}{3}, \frac{1}{4}$ **25.** ____________

26. $\frac{4}{15}, \frac{3}{20}$ **26.** ____________

Add or subtract as indicated. Simplify your answer to lowest terms when appropriate.

27. $\frac{2}{3}+\frac{1}{4}$ **27.** ____________

28. $\frac{4}{15}+\frac{3}{20}$ **28.** ____________

29. $\frac{1}{3}+\frac{2}{7}+\frac{5}{14}$ **29.** ____________

An Application

30. A board measures $27\frac{1}{2}$ inches and needs to be cut into four equal parts. Find the length of each piece. **30.** ____________

Name: **Course/Section:** **Instructor:**

Chapter 1 Introduction to Algebra
1.3 Exponents and Order of Operations

Natural Number Exponents ~ Order of Operations ~ Translating Words to Expressions

STUDY PLAN

Read: Read Section 1.3 on pages 26-32 in your textbook or eText.

Practice: Do your assigned exercises in your ☐ Book ☐ MyMathLab ☐ Worksheets

Review: Keep your corrected assignments in an organized notebook and use them to review for the test.

Key Terms

Exercises 1-3: Use the vocabulary terms listed below to complete each statement. Note that some terms or expressions may not be used. Some terms may be used more than once.

base
parentheses
subtraction
exponent
multiplication
addition
division
absolute value
exponential expression

1. The ________________ b^n, where n is a natural number, means $b^n = b \cdot b \cdot b \cdot \ldots \cdot b$ (with n factors).

2. In the expression b^n, the ________________ is b and the ________________ is n.

3. ORDER OF OPERATIONS
Use the following order of operations. First perform all calculations within ________________ and ________________(s), or above and below the fraction bar.
Evaluate all ________________(s).
Do all ________________ and ________________ from left to right.
Do all ________________ and ________________ from left to right.

Natural Number Exponents

Exercises 1-9: Refer to Examples 1-3 on pages 28-29 in your text and the Section 1.3 lecture video.

Write each product as an exponential expression.

1. $6 \cdot 6 \cdot 6 \cdot 6 \cdot 6$ **1.** ________________

2. $\frac{1}{3} \cdot \frac{1}{3} \cdot \frac{1}{3} \cdot \frac{1}{3}$ **2.** ________________

3. $a \cdot a \cdot a \cdot a \cdot a \cdot a \cdot a$ **3.** ________________

Evaluate each expression.

4. 4^3 **4.** ________________

5. 10^5 **5.** ________________

6. $\left(\frac{2}{3}\right)^3$ **6.** ________________

Use the given base to write each number as an exponential expression. Check your results with a calculator, if one is available.

7. 1000 (base 10) **7.** ________________

8. 32 (base 2) **8.** ________________

9. 81 (base 3) **9.** ________________

Order of Operations

Exercises 10-17: ***Refer to Examples 5-6 on pages 30-31 in your text and the Section 1.3 lecture video.***

Evaluate each expression by hand.

10. $12-7-2$ **10.** ______________

11. $15-(8-1)$ **11.** ______________

12. $4+\frac{9}{3}$ **12.** ______________

13. $\frac{2+1}{7+8}$ **13.** ______________

14. $30-6\cdot 4$ **14.** ______________

15. $6+5\cdot 2-(7-2)$ **15.** ______________

16. $\frac{2+2^3}{25-15}$ **16.** ______________

17. $4\cdot 3^2-(6+2)$ **17.** ______________

Translating Words to Expressions

Exercises 18-21: Refer to Example 7 on page 31 in your text and the Section 1.3 lecture video.

Translate each phrase into a mathematical expression and then evaluate it.

18. Three cubed minus eight **18.** ______________

19. Thirty plus four times two **19.** ______________

20. Four cubed divided by two squared **20.** ______________

21. Forty divided by the quantity ten minus two **21.** ______________

Name: **Course/Section:** **Instructor:**

Chapter 1 Introduction to Algebra
1.4 Real Numbers and the Number Line

Signed Numbers ~ Integers and Rational Numbers ~ Square Roots ~ Real and Irrational Numbers ~ The Number Line ~ Absolute Value ~ Inequality

STUDY PLAN

Read: Read Section 1.4 on pages 34-42 in your textbook or eText.

Practice: Do your assigned exercises in your ☐ Book ☐ MyMathLab ☐ Worksheets

Review: Keep your corrected assignments in an organized notebook and use them to review for the test.

Key Terms

Exercises 1-10: Use the vocabulary terms listed below to complete each statement. Note that some terms or expressions may not be used.

origin
rational number
principal square root
integer
absolute value
opposite
average
additive inverse
irrational number
square root
real number

1. A real number that cannot be expressed by a fraction is a(n) ________________.

2. If a is a positive number, then the ________________ of a, denoted $\sqrt{a}$, is the positive square root of a.

3. The point on the number line associated with the real number 0 is called the ________________.

4. The ________________ of a set of numbers is found by adding the numbers and then dividing by how many numbers there are in the set.

5. A(n) ________________ is any number that can be expressed as a ratio of two integers, $\frac{p}{q}$, where $q \neq 0$.

6. The ________________, or ________________, of a number a is $-a$.

7. The number b is a ________________ of a number a if $b \cdot b = a$.

8. The ________________ of a real number equals its distance on the number line from the origin.

9. The ________________(s) include the natural numbers, zero, and the opposites of the natural numbers.

10. If a number can be represented by a decimal number, then it is a(n) ______________.

Signed Numbers

Exercises 1-4: Refer to Examples 1-2 on page 35 in your text and the Section 1.4 lecture video.

Find the opposite of each expression.

1. 11 — **1.** ______________

2. $-\frac{3}{8}$ — **2.** ______________

3. $-(-5)$ — **3.** ______________

4. Find the additive inverse of $-b$, if $b = -\frac{1}{5}$. — **4.** ______________

Integers and Rational Numbers

Exercises 5-8: Refer to Example 3 on page 36 in your text and the Section 1.4 lecture video.

Classify each number as one or more of the following: natural number, whole number, integer, or rational number.

5. $\frac{15}{3}$ **5.** ________________

6. -4 **6.** ________________

7. 0 **7.** ________________

8. $-\frac{8}{3}$ **8.** ________________

Square Roots

Exercises 9-11: Refer to Example 4 on page 37 in your text and the Section 1.4 lecture video.

Evaluate each square root. Approximate your answer to three decimal places when appropriate.

9. $\sqrt{64}$ **9.** ________________

10. $\sqrt{400}$ **10.** ________________

11. $\sqrt{6}$ **11.** ________________

Real and Irrational Numbers

Exercises 12-16: Refer to Examples 5-6 on pages 38-39 in your text and the Section 1.4 lecture video.

Classify each number as one or more of the following: natural number, whole number, integer, rational number, or irrational number.

12. $-\sqrt{7}$ **12.** ______________

13. -2 **13.** ______________

14. $\frac{9}{14}$ **14.** ______________

15. $\sqrt{25}$ **15.** ______________

16. The table lists the height (in inches) of four students. Find the average height. Is the result a natural number, a rational number, or an irrational number? **16.** ______________

____

Student	1	2	3	4
Height (in inches)	71	68	72	65

The Number Line

Exercises 17-19: Refer to Example 7 on page 39 in your text and the Section 1.4 lecture video.

Plot each real number on a number line.

17. $\frac{3}{4}$

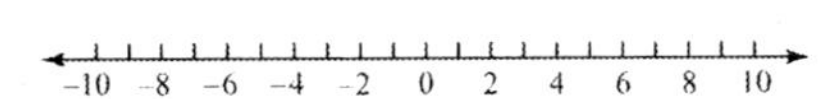

18. $\sqrt{5}$

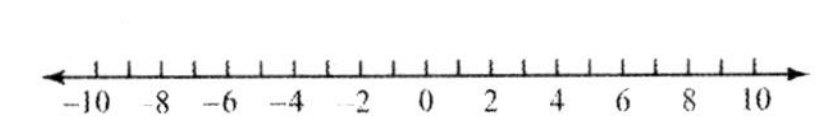

19. $-\pi$

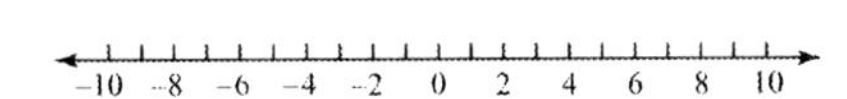

Absolute Value

Exercises 20-22: Refer to Example 8 on page 40 in your text and the Section 1.4 lecture video.

Evaluate each expression.

20. $|6.7|$ **20.** ________________

21. $|-4|$ **21.** ________________

22. $|3-10|$ **22.** ________________

Inequality

Exercise 23: Refer to Example 9 on page 41 in your text and the Section 1.4 lecture video.

23. List the following numbers from least to greatest. Then plot these numbers on a number line. **23.** ________________

$\pi, -\sqrt{3}, 0, 2.2, -3$

−6 −4 −2 0 2 4 6

Name: **Course/Section:** **Instructor:**

Chapter 1 Introduction to Algebra
1.5 Addition and Subtraction of Real Numbers

Addition of Real Numbers ~ Subtraction of Real Numbers ~ Applications

STUDY PLAN

Read: Read Section 1.5 on pages 45-49 in your textbook or eText.

Practice: Do your assigned exercises in your ☐ Book ☐ MyMathLab ☐ Worksheets

Review: Keep your corrected assignments in an organized notebook and use them to review for the test.

Key Terms

Exercises 1-7: Use the vocabulary terms listed below to complete each statement. Note that some terms or expressions may not be used.

sum	**greater than**
addends	**difference**
less than	**less than or equal to**
greater than or equal to	**approximately equal**
$a > b$	$a \geq b$
$a \leq b$	$a < b$

1. We say that a is ________________ b, denoted ________________, if either $a < b$ or $a = b$ is true.

2. The answer to a subtraction problem is the ________________.

3. If a real number a is located to the right of a real number b on the number line, we say that a is ________________ b and we write ________________.

4. The symbol $\approx$ is used to represent ________________.

5. We say that a is ________________ b, denoted ________________, if either $a > b$ or $a = b$ is true.

6. In an addition problem the two numbers added are called ________________, and the answer is called the ________________.

7. If a real number a is located to the left of a real number b on the number line, we say that a is ________________ b and we write ________________.

Addition of Real Numbers

Exercises 1-10: Refer to Examples 1-3 on pages 45-46 in your text and the Section 1.5 lecture video.

Find the opposite of each number and calculate the sum of the number and its opposite.

1. 32 **1.** ______________

____

2. $-\sqrt{3}$ **2.** ______________

____

3. $\frac{5}{6}$ **3.** ______________

____

Find each sum by hand.

4. $-4+(-9)$ **4.** ______________

5. $-\frac{3}{4}+\frac{4}{5}$ **5.** ______________

6. $5.1+(-9.4)$ **6.** ______________

Add visually, using the symbols $\cap$ and $\cup$.

7. $5+3$ **7.** ______________

8. $-7+4$ **8.** ______________

9. $2+(-6)$ **9.** ______________

10. $-7+(-3)$ **10.** ______________

Subtraction of Real Numbers

Exercises 11-17: Refer to Examples 4-5 on pages 47-48 in your text and the Section 1.5 lecture video.

Find each difference by hand.

11. $8-28$ **11.** ____________

12. $-3-9$ **12.** ____________

13. $-6.7-(-4.3)$ **13.** ____________

14. $-\frac{2}{3}-\left(-\frac{4}{5}\right)$ **14.** ____________

Evaluate each expression by hand.

15. $8-4-(-2)+3$ **15.** ____________

16. $-\frac{5}{6}+\frac{1}{3}-\left(-\frac{3}{4}\right)$ **16.** ____________

17. $-3.1+6.9-10.2$ **17.** ____________

Applications

Exercises 18-19: Refer to Examples 6-7 on page 49 in your text and the Section 1.5 lecture video.

18. In a nine-month period in Punta Arenas, at the southern tip of Chile, the temperature, with the wind chill factor, ranged from $-120°F$ to $52°F$. Find the difference between these temperatures.

18. ______________

19. The initial balance in a checking account is $315. Find the final balance if the following represents a list of withdrawals and deposits: –$65, –$40, $140, and –$85.

19. ______________

Name: Course/Section: Instructor:

Chapter 1 Introduction to Algebra
1.6 Multiplication and Division of Real Numbers

Multiplication of Real Numbers ~ Division of Real Numbers ~ Applications

STUDY PLAN

Read: Read Section 1.6 on pages 51-58 in your textbook or eText.

Practice: Do your assigned exercises in your ☐ Book ☐ MyMathLab ☐ Worksheets

Review: Keep your corrected assignments in an organized notebook and use them to review for the test.

Key Terms

Exercises 1-4: Use the vocabulary terms listed below to complete each statement. Note that some terms or expressions may not be used.

quotient
reciprocal
divisor
positive
dividend
negative
multiplicative inverse

1. The product or quotient of two numbers with unlike signs is a(n) ________________ number.

2. The ________________, or ________________, of a real number a is $\frac{1}{a}$.

3. The product or quotient of two numbers with like signs is a(n) ________________ number.

4. In the division problem $20 \div 4 = 5$, the number 20 is the ________________, 4 is the ________________, and 5 is the ________________.

Multiplication of Real Numbers

Exercises 1-8: Refer to Examples 1-2 on pages 52-53 in your text and the Section 1.6 lecture video.

Find each product by hand.

1. $-6 \cdot 4$ **1.** ________________

2. $\frac{3}{4} \cdot \frac{2}{5}$ **2.** ________________

3. $-3.2(-1.5)$ **3.** ________________

4. $(1.2)(-3)(-4)(-5)$ **4.** ________________

Evaluate each expression by hand.

5. $(-5)^2$ **5.** ________________

6. -5^2 **6.** ________________

7. $(-4)^3$ **7.** ________________

8. -4^3 **8.** ________________

Division of Real Numbers

Exercises 9-20: Refer to Examples 3-6 on pages 54-56 in your text and the Section 1.6 lecture video.

Evaluate each expression by hand.

9. $-15 \div \frac{1}{3}$ **9.** ______________

10. $\frac{\frac{2}{3}}{6}$ **10.** ______________

11. $-\frac{-12}{-32}$ **11.** ______________

12. $0 \div (-4)$ **12.** ______________

Convert each measurement to a decimal number.

13. $\frac{5}{6}$-inch nail **13.** ______________

14. $\frac{7}{16}$-inch radius **14.** ______________

15. $1\frac{1}{8}$-cup sugar **15.** ______________

Convert each decimal number to a fraction in lowest terms.

16. 0.08 **16.** ______________

17. 0.275 **17.** ______________

18. 0.005 **18.** ______________

Use a calculator to evaluate each expression. Express your answer as a decimal and as a fraction.

19. $\frac{5}{6}-\frac{1}{2}+\frac{4}{3}$ **19.** ______________

20. $\left(\frac{2}{3}\cdot\frac{5}{8}\right)\div\frac{4}{9}$ **20.** ______________

Applications

Exercises 21-22: Refer to Examples 7-8 on pages 56-57 in your text and the Section 1.6 lecture video.

21. Lotto winnings of $1.2 million were divided among three people. Because he purchased the ticket and chose the numbers, Sam received $\frac{7}{15}$ of the total winnings, while his two friends divided the remaining amount.

(a) What amount did Sam receive? **21.(a)**______________

(b) What amount did each of Sam's friends receive? **(b)**______________

22. A student planned to spend $\frac{19}{250}$ of his education budget for textbooks. Write this fraction as a decimal. **22.** ______________

Name: ____________ **Course/Section:** ____________ **Instructor:** ____________

Chapter 1 Introduction to Algebra
1.7 Properties of Real Numbers

Commutative Properties ~ Associative Properties ~ Distributive Properties ~ Identity and Inverse Properties ~ Mental Calculations

STUDY PLAN

Read: Read Section 1.7 on pages 60-68 in your textbook or eText.

Practice: Do your assigned exercises in your ☐ Book ☐ MyMathLab ☐ Worksheets

Review: Keep your corrected assignments in an organized notebook and use them to review for the test.

Key Terms

Exercises 1-9: Use the vocabulary terms listed below to complete each statement. Note that some terms or expressions may not be used.

additive inverse property
distributive property
associative property for addition
identity property of 1
commutative property for addition
commutative property for multiplication
identity property of 0
multiplicative inverse property
associative property for multiplication

1. The ________________ states that for any real number a, $a+0=0+a=a$.

2. The ________________ states that for any real numbers a, b and c, $(a+b)+c=a+(b+c)$.

3. The ________________ states that for any nonzero real number a, $a\cdot\frac{1}{a}=1$ and $\frac{1}{a}\cdot a=1$.

4. The ________________ states that for any real numbers a, b and c, $a(b+c)=ab+ac$ and $a(b-c)=ab-ac$.

5. The ________________ states that for any real number, a and b, $a\cdot b=b\cdot a$.

6. The ________________ states that if any number a is multiplied by 1, the result is a.

7. The ________________ states that for any real numbers a, b and c, $(a \cdot b) \cdot c = a \cdot (b \cdot c)$.

8. The ________________ states that for any real number a, $a + (-a) = 0$ and $-a + a = 0$.

9. The ________________ states that two numbers a and b can be added in any order and the result will be the same.

Commutative Properties

Exercises 1-2: Refer to Example 1 on page 61 in your text and the Section 1.7 lecture video.

Use a commutative property to rewrite each expression.

1. $4 + 20$ **1.** ________________

2. $x \cdot 9$ **2.** ________________

Associative Properties

Exercises 3-7: Refer to Examples 2-3 on page 62 in your text and the Section 1.7 lecture video.

Use an associative property to rewrite each expression.

3. $(2 + 4) + 7$ **3.** ________________

4. $a(bc)$ **4.** ________________

State the property that each equation illustrates.

5. $5 + (1 + x) = (5 + 1) + x$ **5.** ________________

6. $x \cdot y = y \cdot x$ **6.** ________________

7. $2 + ab = ab + 2$ **7.** ________________

Distributive Properties

Exercises 8-18: Refer to Examples 4-6 on pages 64-65 in your text and the Section 1.7 lecture video.

Apply a distributive property to each expression.

8. $4(a-5)$ **8.** ______________

9. $-3(b+8)$ **9.** ______________

10. $-(x-9)$ **10.** ______________

11. $14-(y+3)$ **11.** ______________

Use the distributive property to insert parentheses in the expression and then simplify the result.

12. $7x+4x$ **12.** ______________

13. $2a-10a$ **13.** ______________

14. $-6y+3y$ **14.** ______________

State the property or properties illustrated by each equation.

15. $(3+x)+7=x+10$ **15.** ______________

16. $4(12a)=48a$ **16.** ______________

17. $-3(x-9)=-3x+27$ **17.** ______________

18. $2(b+a)=2a+2b$ **18.** ______________

Identity and Inverse Properties

Exercises 19-22: Refer to Example 7 on page 66 in your text and the Section 1.7 lecture video.

State the property or properties illustrated by each equation.

19. $\frac{12}{32}=\frac{3}{8}\cdot\frac{4}{4}=\frac{3}{8}$ **19.** ________________

20. $-7+0=-7$ **20.** ________________

21. $\frac{1}{3}\cdot 3a=1\cdot a=a$ **21.** ________________

22. $2+(-2)+x=0+x=x$ **22.** ________________

Mental Calculations

Exercises 23-27: Refer to Examples 8-9 on page 67 in your text and the Section 1.7 lecture video.

Use properties of real numbers to calculate each expression mentally.

23. $\frac{4}{3}\cdot 2\cdot\frac{3}{4}\cdot\frac{1}{2}$ **23.** ________________

24. $12+5+8+25$ **24.** ________________

25. $3\cdot 65$ **25.** ________________

26. $434+99$ **26.** ________________

27. A tank is 40 feet long, 25 feet wide, and 5 feet deep. The volume V of the tank is found by multiplying 40, 25 and 5. Use the commutative and associative properties for multiplication to calculate the volume of the tank mentally. **27.** ________________

Name: **Course/Section:** **Instructor:**

Chapter 1 Introduction to Algebra
1.8 Simplifying and Writing Algebraic Expressions

Terms ~ Combining Like Terms ~ Simplifying Expressions ~ Writing Expressions

STUDY PLAN

Read: Read Section 1.8 on pages 71-76 in your textbook or eText.

Practice: Do your assigned exercises in your ☐ Book ☐ MyMathLab ☐ Worksheets

Review: Keep your corrected assignments in an organized notebook and use them to review for the test.

Key Terms

Exercises 1-5: Use the vocabulary terms listed below to complete each statement. Note that some terms or expressions may not be used.

term
multiplicative identity
like term
coefficient
additive identity

1. The number 0 is called the ___________________.

2. A(n) ___________________ is a number, a variable, or a product of numbers and variables raised to natural number powers.

3. The ___________________ of a term is the number that appears in the term.

4. If two terms contain the same variables raised to the same powers, we call them ___________________(s).

5. The number 1 is called the ___________________.

Terms

Exercises 1-4: Refer to Example 1 on page 72 in your text and the Section 1.8 lecture video.

Determine whether each expression is a term. If it is a term, identify its coefficient.

1. -2 **1.** ________

2. $4x$ **2.** ________

3. $3x-7y$ **3.** ________

4. $6a^3$ **4.** ________

Combining Like Terms

Exercises 5-11: Refer to Examples 2-3 on page 73 in your text and the Section 1.8 lecture video.

Determine whether the terms are like or unlike.

5. $-2a, -2b$ **5.** ________

6. $4x^2, -9x^2$ **6.** ________

7. $3z, 5z^2$ **7.** ________

8. $-3m, 8m$ **8.** ________

Combine terms in each expression, if possible.

9. $-4x+\frac{1}{2}x$ **9.** ________

10. $7y^2-y^2$ **10.** ________

11. $-a^2+4a$ **11.** ________

Simplifying Expressions

Exercises 12-19: Refer to Examples 4-6 on pages 74-75 in your text and the Section 1.8 lecture video.

Simplify each expression.

12. $3-x-8+4x$ **12.** ____________

13. $8y-(y+13)$ **13.** ____________

14. $\dfrac{-2.7a}{-2.7}$ **14.** ____________

15. $10-3(b+4)$ **15.** ____________

16. $5x^2-x+3x^2+x$ **16.** ____________

17. $7t^3-t^3-12t^3$ **17.** ____________

18. $\dfrac{12z-9}{3}$ **18.** ____________

19. Simplify the expression $7a+8-2a-13$. **19.** ____________

Writing Expressions

Exercise 20: Refer to Example 7 on page 76 in your text and the Section 1.8 lecture video.

20. A street has a constant width w and comprises several short sections having lengths 450, 600, 520, and 700 feet.

(a) Write and simplify an expression that gives the square footage of the street. **20. (a)**____________

(b) Find the area of the street if its width is 48 feet. **(b)**____________

Name: **Course/Section:** **Instructor:**

Chapter 2 Linear Equations and Inequalities
2.1 Introduction to Equations

Basic Concepts ~ Equations and Solutions ~ The Addition Property of Equality ~ The Multiplication Property of Equality

STUDY PLAN

Read: Read Section 2.1 on pages 90-97 in your textbook or eText.

Practice: Do your assigned exercises in your ☐ Book ☐ MyMathLab ☐ Worksheets

Review: Keep your corrected assignments in an organized notebook and use them to review for the test.

Key Terms

Exercises 1-5: Use the vocabulary terms listed below to complete each statement. Note that some terms or expressions may not be used.

solution
equivalent
solution set
addition property of equality
multiplication property of equality

1. The ________________ states that if a, b, and c are real numbers, then $a = b$ is equivalent to $a + c = b + c$.

2. The set of all solutions to an equation is called the ________________.

3. The ________________ states that multiplying each side of an equation by the same nonzero number results in an equivalent equation.

4. ________________ equations are equations that have the same solution set.

5. Each value of the variable that makes an equation true is called a(n) ________________ to the equation.

The Addition Property of Equality

Exercises 1-4: Refer to Examples 1-2 on pages 92-93 in your text and the Section 2.1 lecture video.

Solve each equation.

1. $x+12=6$ **1.** ________________

2. $t-2=9$ **2.** ________________

3. $\frac{2}{3}=-\frac{1}{2}+x$ **3.** ________________

4. Solve the equation $-6+y=-5$ and then check the solution. **4.** ________________

The Multiplication Property of Equality

Exercises 5-9: Refer to Examples 3-5 on pages 94-96 in your text and the Section 2.1 lecture video.

Solve each equation.

5. $\frac{1}{3}x=-3$ **5.** ________________

6. $-3t=15$ **6.** ________________

7. $6=\frac{2}{3}x$ **7.** ________________

8. Solve the equation $\frac{3}{4}=-\frac{3}{2}z$ and then check the solution. **8.** ________________

9. A 2009 study concluded that 24 cubic miles of Alaskan glacier ice are lost each year.

(a) Write a formula that gives the number of cubic miles of glacier ice lost in x years. **9. (a)**________________

(b) At this rate, how many years will it take for 600 cubic miles of glacier ice to melt? **(b)**________________

Name: Course/Section: Instructor:

Chapter 2 Linear Equations and Inequalities
2.2 Linear Equations

Basic Concepts ~ Solving Linear Equations ~ Applying the Distributive Property ~ Clearing Fractions and Decimals ~ Equations with No Solutions or Infinitely Many Solutions

STUDY PLAN

Read: Read Section 2.2 on pages 99-108 in your textbook or eText.

Practice: Do your assigned exercises in your ☐ Book ☐ MyMathLab ☐ Worksheets

Review: Keep your corrected assignments in an organized notebook and use them to review for the test.

Key Terms

Exercises 1-6: Use the vocabulary terms listed below to complete each statement. Note that some terms or expressions may not be used.

identity	**linear equation**
no solutions	**contradiction**
one solution	**infinitely many solutions**

1. A(n) ________________ in one variable is an equation that can be written in the form $ax+b=0$, where a and b are constants with $a \neq 0$.

2. If the process of solving an equation results in a contradiction, such as $0=4$ or $5=1$, the equation has ________________.

3. If the process of solving an equation results in an identity, such as $0=0$ or $-2=-2$, the equation has ________________.

4. An equation that is always false is called a(n) ________________.

5. If the process of solving an equation results in a statement which is true for only one value, such as $x=3$ or $x=-7$, the equation has ________________.

6. An equation that is always true is called a(n) ________________.

Basic Concepts

Exercises 1-4: Refer to Example 1 on pages 100-101 in your text and the Section 2.2 lecture video.

Determine whether the equation is linear. If the equation is linear, give values for a and b that result when the equation is written in the form $ax+b=0$.

1. $3x-2=0$ **1.** ________________

2. $4x^2-3=0$ **2.** ________________

3. $\frac{1}{2}=7$ **3.** ________________

4. $\frac{2}{3}x-6=0$ **4.** ________________

Solving Linear Equations

Exercises 5-9: Refer to Examples 2-4 on pages 101-104 in your text and the Section 2.2 lecture video.

5. Complete the table for the given values of x. Then solve the equation $-2x-5=-1$. **5.** ________________

x	-3	-2	-1	0	1	2	3
$-2x-5$							

Solve each linear equation. Check the answer.

6. $4x-7=0$ **6.** ________________

7. $\frac{1}{3}x+4=2$ **7.** ________________

8. $6x-1=2x+9$ **8.** ________________

9. The number of Internet users I in millions during year x can be approximated by the formula $I = 241x - 482{,}440$, where $x \geq 2007$. Estimate the year when there were 1970 million (1.97 billion) Internet users.

9. ______________

Applying the Distributive Property

Exercises 10-11: Refer to Example 5 on pages 104-105 in your text and the Section 2.2 lecture video.

Solve each linear equation. Check the answer.

10. $3(x+5)+x=0$

10. ______________

11. $3(2z-7)+4=4(z-1)$

11. ______________

Clearing Fractions and Decimals

Exercises 12-15: Refer to Examples 6-7 on pages 105-106 in your text and the Section 2.2 lecture video.

Solve each linear equation.

12. $\frac{4}{3}=\frac{1}{6}-\frac{5}{4}b$

12. ______________

13. $\frac{2}{3}x-\frac{3}{4}=\frac{1}{2}-x$

13. ______________

14. $2.5x-1.4=4.2$

14. ______________

15. $5.1-2.2y=y+3.7$

15. ______________

Equations with No Solutions or Infinitely Many Solutions

Exercises 16-18: Refer to Example 8 on page 107 in your text and the Section 2.2 lecture video.

Determine whether the equation has no solutions, one solution, or infinitely many solutions.

16. $3x-2(x-2)=x+4$ **16.** ________________

17. $3x=5x+2(x+4)$ **17.** ________________

18. $6x+7=2(3x-2)+2$ **18.** ________________

Name: **Course/Section:** **Instructor:**

Chapter 2 Linear Equations and Inequalities
2.3 Introduction to Problem Solving

Steps for Solving a Problem ~ Percent Problems ~ Distance Problems ~ Other Types of Problems

STUDY PLAN

Read: Read Section 2.3 on pages 111-120 in your textbook or eText.

Practice: Do your assigned exercises in your ☐ Book ☐ MyMathLab ☐ Worksheets

Review: Keep your corrected assignments in an organized notebook and use them to review for the test.

Key Terms

Exercises 1-7: Use the vocabulary terms listed below to complete each statement. Note that some terms or expressions may not be used.

is	**less**	**per**
plus	**double**	**add**
subtract	**total**	**sum**
decimal number	**results in**	**gives**
quotient	**fraction**	**minus**
equals	**product**	**increase**
decrease	**divide**	**fewer**
multiply	**times**	**triple**
more	**difference**	**percent change**
twice	**divided by**	**is the same as**

1. If a quantity changes from an old amount to a new amount , the ________________ is given by $\frac{\text{new amount} - \text{old amount}}{\text{old amount}} \cdot 100.$

2. The expression $x\%$ represents the ________________ $\frac{x}{100}$ or the ________________ $x \cdot 0.01$.

3. List six words/phrases that are associated with the math symbol $+$.

4. List six words/phrases that are associated with the math symbol $-$.

5. List six words/phrases that are associated with the math symbol $\cdot$.

6. List four words/phrases that are associated with the math symbol $\div$.

7. List five words/phrases that are associated with the math symbol $=$.

Steps for Solving a Problem

Exercises 1-5: Refer to Examples 1-3 on pages 112-113 in your text and the Section 2.3 lecture video.

Translate the sentence into an equation using the variable x. Then solve the resulting equation.

1. Four times a number plus 5 is equal to 17. **1.** ______________

2. The sum of one-third a number and 7 is 1. **2.** ______________

3. Fifteen is 3 less than twice a number. **3.** ______________

4. The sum of three consecutive natural numbers is 54. Find the three numbers. **4.** ______________

5. The population of a small town is 34,000. This is 2000 more than twice the population of a neighboring town. Find the population of the neighboring town. **5.** ______________

Percent Problems

Exercises 6-14: Refer to Examples 4-8 on pages 114-116 in your text and the Section 2.3 lecture video.

Convert each percentage to fraction and decimal notation.

6. 33% **6.** ______________

7. 2.5% **7.** ______________

8. 0.8% **8.** ______________

Convert each real number to a percentage.

9. 0.137 **9.** ______________

10. $\frac{1}{8}$ **10.** ______________

11. 1.61 **11.** ______________

12. From 2005 to 2010, the cost of tuition increased from \$130 to \$156 per credit. Calculate the percent increase in tuition cost from 2005 to 2010. **12.** ______________

13. In 2010, an office manager received a salary of \$57,500. With the downturn in the economy in 2011, he was forced to take a 3% pay cut. Calculate his salary after the cut. **13.** ______________

14. In 2050, about 16% of the world's population, or 1.5 billion people, will be older than 65. Find the estimated population of the world in 2050. **14.** ______________

Distance Problems

Exercises 15-16: Refer to Examples 9-10 on pages 116-117 in your text and the Section 2.3 lecture video.

15. A pilot flies a plane at a constant speed for 3 hours and 20 minutes, traveling 1900 miles. Find the speed of the plane in miles per hour.

15. ________________

16. A marathoner jogs at two speeds, covering a distance of 15 miles in 2 hours. If she runs one-half hour at 6 miles per hour, find the second speed.

16. ________________

Other Types of Problems

Exercises 17-18: Refer to Examples 11-12 on pages 118-119 in your text and the Section 2.3 lecture video.

17. A solution contains 5% salt. How much pure water should be added to 20 ounces of the solution to dilute it to a 2% solution?

17. ________________

18. A student takes out a loan for a limited amount of money at 9% interest and then must pay 11% for any additional money. If the student borrows $3500 less at 11% than at 9%, then the total interest for one year is $615. How much does the student borrow at each rate?

18. ________________

Name: **Course/Section:** **Instructor:**

Chapter 2 Linear Equations and Inequalities
2.4 Formulas

Formulas from Geometry ~ Solving for a Variable ~ Other Formulas

STUDY PLAN

Read: Read Section 2.4 on pages 124-133 in your textbook or eText.

Practice: Do your assigned exercises in your ☐ Book ☐ MyMathLab ☐ Worksheets

Review: Keep your corrected assignments in an organized notebook and use them to review for the test.

Key Terms

Exercises 1-8: Use the vocabulary terms listed below to complete each statement. Note that some terms or expressions may not be used. Some terms may be used more than once.

perimeter
area
degree
circumference
volume

1. The ________________ of a circle with radius r is given by πr^2.

2. If a rectangle has length l and width w, then its ________________ is given by lw.

3. The ________________ of a cylinder having radius r and height h is given by $\pi r^2 h$.

4. The perimeter of a circle is called its ________________.

5. If a rectangle has length l and width w, then its ________________ is given by $2l + 2w$.

6. A(n) ________________ is $\frac{1}{360}$ of a revolution.

7. If a triangle has base b and height h, then its ________________ is given by $\frac{1}{2}bh$.

8. The ________________ of a circle with radius r is given by $2\pi r$.

Formulas from Geometry

Exercises 1-6: Refer to Examples 1-6 on pages 125-128 in your text and the Section 2.4 lecture video.

1. A residential lot is shown in the figure. It comprises a rectangular region and an adjacent triangular region.

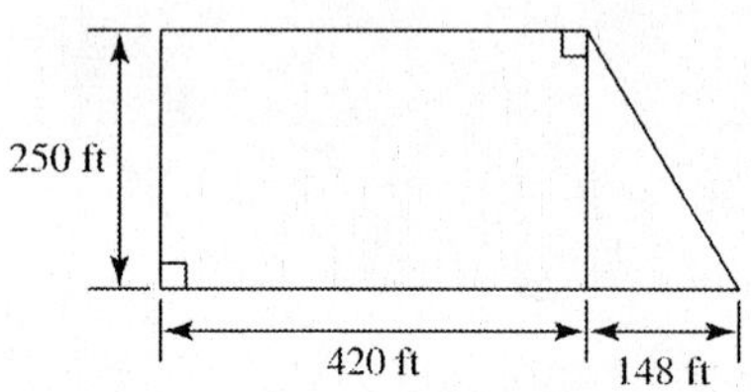

(a) Find the area of this lot. **1.(a)**____________

(b) An acre contains 43,560 square feet. How many acres are there in this lot? **(b)**____________

2. In a triangle, the two larger angles are equal in measure and each is twice the measure of the smallest angle. Find the measure of each angle. **2.** ____________

____

____

3. A circle has a diameter of 15 centimeters. Find its circumference and area. **3.** ____________

____

4. Find the area of the trapezoid shown in the figure. **4.** ____________

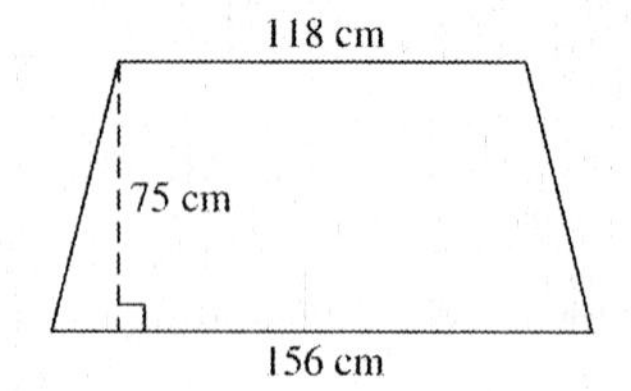

5. Find the volume and surface area of a box with length 12 meters, width 5 meters, and height 2 meters.

5. ________________
____

6. A cylindrical soup can has a radius $1\frac{1}{4}$ of inches and a height of 4 inches.

(a) Find the volume of the can.

6.(a)________________

(b) If 1 cubic inch equals 0.554 fluid ounce, find the number of fluid ounces in the can.

(b)________________

Solving for a Variable

Exercises 7-9: Refer to Examples 7-8 on pages 129-130 in your text and the Section 2.4 lecture video.

7. The perimeter of a rectangle is given by $P = 2l + 2w$.

(a) Solve the formula for l.

7.(a)________________

(b) A rectangle has perimeter $P = 26$ inches and width $w = 5$ inches. Find l.

(b)________________

Solve each equation for the indicated variable.

8. $A = \frac{1}{2}(a+b)h$ for h

8. ________________

9. $xy + yz = xz$ for x

9. ________________

Other Formulas

Exercises 10-11: Refer to Examples 9-10 on pages 130-131 in your text and the Section 2.4 lecture video.

10. A student has earned 12 credits of A, 30 credits of B, 22 credits of C, 4 credits of D, and 4 credits of F. Calculate the student's GPA to the nearest hundredth.

10. ______________

11. To convert Celsius degrees C to Fahrenheit degrees F, the formula $F = \frac{9}{5}C + 32$ can be used.

(a) Solve the formula for C to find a formula that converts Fahrenheit degrees to Celsius degrees.

11. (a)______________

(b) If the temperature is $77°\text{F}$, find the equivalent Celsius temperature.

(b)______________

Name: Course/Section: Instructor:

Chapter 2 Linear Equations and Inequalities
2.5 Linear Inequalities

Solutions and Number Line Graphs ~ The Addition Property of Inequalities ~ The Multiplication Property of Inequalities ~ Applications

STUDY PLAN

Read: Read Section 2.5 on pages 136-145 in your textbook or eText.

Practice: Do your assigned exercises in your ☐ Book ☐ MyMathLab ☐ Worksheets

Review: Keep your corrected assignments in an organized notebook and use them to review for the test.

Key Terms

Exercises 1-5: Use the vocabulary terms listed below to complete each statement. Note that some terms or expressions may not be used.

solution
solution set
interval notation
linear inequality
set-builder notation

1. A solution in the form $(-3,4]$ is written in ________________.

2. The set of all solutions to an inequality is called the ________________.

3. A solution in the form $\{x|x \le -2\}$ is written in ________________.

4. A(n) ________________ results whenever the equals sign in a linear equation is replaced with any one of the symbols $<, \le, >,$ or $\ge$.

5. A(n) ________________ to an inequality is a value of the variable that makes the statement true.

Solutions and Number Line Graphs

Exercises 1-13: Refer to Examples 1-4 on pages 137-138 in your text and the Section 2.5 lecture video.

Use a number line to graph the solution set to each inequality.

1. $x < 0$

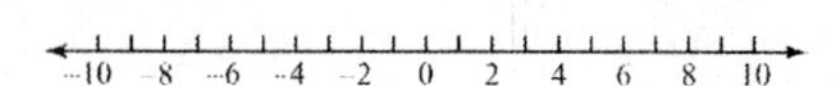

2. $x \leq 0$

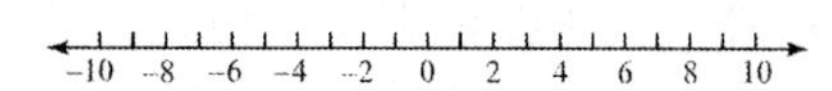

3. $x > 2$

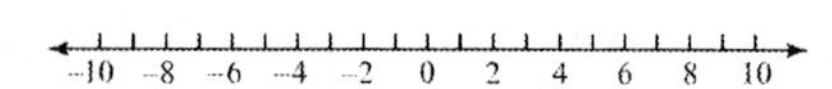

4. $x \leq 5$

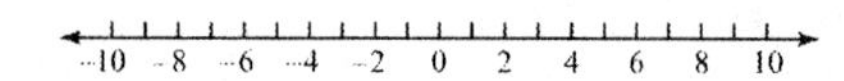

Write the solution set to each inequality in interval notation.

5. $x < 7$ **5.** ________________

6. $y \geq -4$ **6.** ________________

7. $a > 0$ **7.** ________________

Determine whether the given value of x is a solution to the inequality.

8. $2x - 7 \leq 5,\ x = 3$ **8.** ________________

9. $6 - 4x > 10,\ x = -1$ **9.** ________________

In the table, the expression $-3x+5$ has been evaluated for several values of x. Use the table to determine any solutions to each equation or inequality.

x	-2	-1	0	1	2	3	4
$-3x+5$	11	8	5	2	-1	-4	-7

10. $-3x+5=5$

10. ______________

11. $-3x+5>5$

11. ______________

12. $-3x+5\geq 5$

12. ______________

13. $-3x+5<5$

13. ______________

The Addition Property of Inequalities

Exercises 14-16: Refer to Examples 5-6 on pages 139-140 in your text and the Section 2.5 lecture video.

Solve each inequality. Then graph the solution set.

14. $x+3\leq -6$

−10 −8 −6 −4 −2 0 2 4 6 8 10

14. ______________

15. $4+2x>7+x$

−10 −8 −6 −4 −2 0 2 4 6 8 10

15. ______________

16. $6-\frac{1}{2}x\geq 4+\frac{1}{2}x$

−10 −8 −6 −4 −2 0 2 4 6 8 10

16. ______________

The Multiplication Property of Inequalities

Exercises 17-21: Refer to Examples 7-8 on pages 141-143 in your text and the Section 2.5 lecture video.

Solve each inequality. Then graph the solution set.

17. $-2x \le 14$

17. 

18. $-3 > -\frac{1}{3}x$

18. ________________

Solve each inequality. Write the solution set in set-builder notation.

19. $3x - 8 < -6$

19. ________________

20. $-6 + 6x \le 5x + 2$

20. ________________

21. $0.3(2x - 1) > -2.3x - 9$

21. ________________

Applications

Exercises 22-26: Refer to Examples 9-11 on pages 143-144 in your text and the Section 2.5 lecture video.

Translate each phrase to an inequality. Let the variable be x.

22. A number that is less than -10 **22.** ______________

23. A temperature that is at most $90°F$ **23.** ______________

24. A grade point average that is at least 3.5 **24.** ______________

25. If the ground temperature is $81°F$, then the air temperature x miles high is given by the formula $T = 81 - 19x$. Determine the altitudes at which the air temperature is less than $5°F$. **25.** ______________

26. To manufacture a certain product, a company incurs a one-time fixed cost of \$4200 plus a per-item cost of \$180. The selling price of the item is \$240.

(a) Write a formula that gives the cost C of producing x items. **26. (a)**______________

(b) Write a formula that gives the revenue R from selling x items. **(b)**______________

(c) Profit equals revenue minus cost. Write a formula that calculates the profit P from producing and selling x items. **(c)**______________

(d) How many items must be sold to yield a positive profit? **(d)**______________

Name: _______________ **Course/Section:** _______________ **Instructor:** _______________

Chapter 3 Graphing Equations
3.1 Introduction to Graphing

Tables and Graphs ~ The Rectangular Coordinate System ~ Scatterplots and Line Graphs

STUDY PLAN

Read: Read Section 3.1 on pages 159-165 in your textbook or eText.

Practice: Do your assigned exercises in your ☐ Book ☐ MyMathLab ☐ Worksheets

Review: Keep your corrected assignments in an organized notebook and use them to review for the test.

Key Terms

Exercises 1-8: Use the vocabulary terms listed below to complete each statement. Note that some terms or expressions may not be used.

***y*-axis**	**scatterplot**
origin	***y*-coordinate**
x-axis	**ordered pair**
line graph	***x*-coordinate**
quadrants	**rectangular coordinate system**

1. In the xy-plane, a point is graphed as a(n) _______________ (x, y).

2. If consecutive data points are connected with line segments, then the resulting graph is called a(n) _______________.

3. When plotting in the xy-plane, the first value in an ordered pair is called the _______________ and the second value is called the _______________.

4. The point of intersection of the axes is called the _______________ and is associated with zero on each axis.

5. If distinct points are plotted in the xy-plane, then the resulting graph is called a(n) _______________.

6. The axes divide the xy-plane into four regions called _______________, which are numbered I, II, III, and IV counterclockwise.

7. One common way to graph data is to use the _______________, or xy-plane.

8. In the xy-plane the horizontal axis is the _______________, and the vertical axis is the _______________.

The Rectangular Coordinate System

Exercises 1-4: Refer to Examples 1-2 on pages 160-161 in your text and the Section 3.1 lecture video.

Plot the following ordered pairs on the same xy-plane. State the quadrant in which each point is located, if possible.

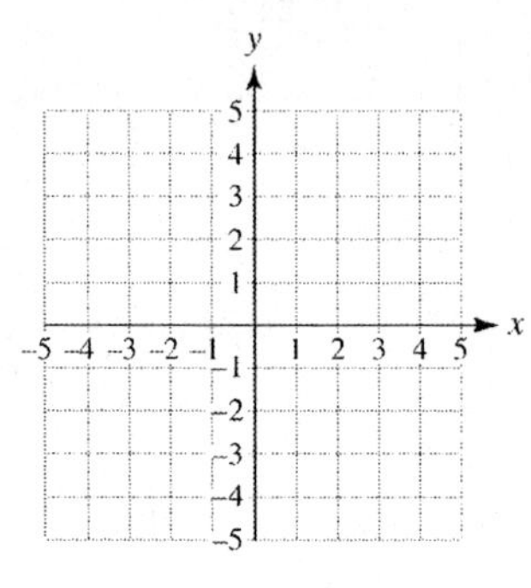

1. $(2,4)$ **1.** ____________

2. $(-3,-1)$ **2.** ____________

3. $(0,-2)$ **3.** ____________

4. The figure shows the average number of hours of video posted to YouTube every minute during selected months. Use the graph to estimate the number of hours of video posted to YouTube every minute in December 2008 and June 2010. **4.** ____________

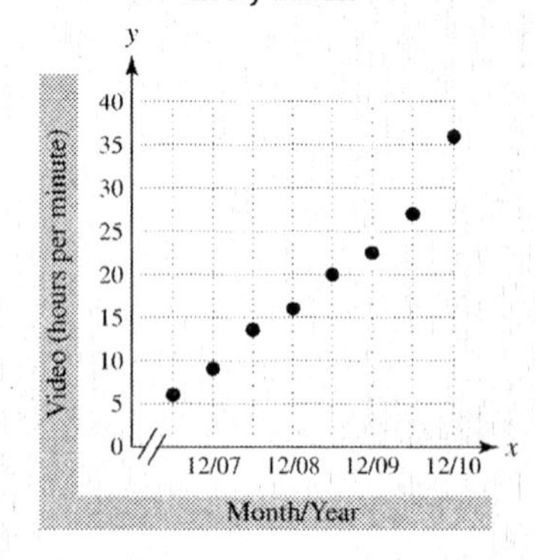

Scatterplots and Line Graphs

Exercises 5-7: Refer to Examples 3-5 on pages 162-164 in your text and the Section 3.1 lecture video.

5. The table lists the average price of a gallon of milk for selected years. Make a scatterplot of the data.

Year	2006	2007	2008	2009	2010	2011
Cost (per gal)	$2.99	$3.45	$2.65	$2.69	$2.79	$3.39

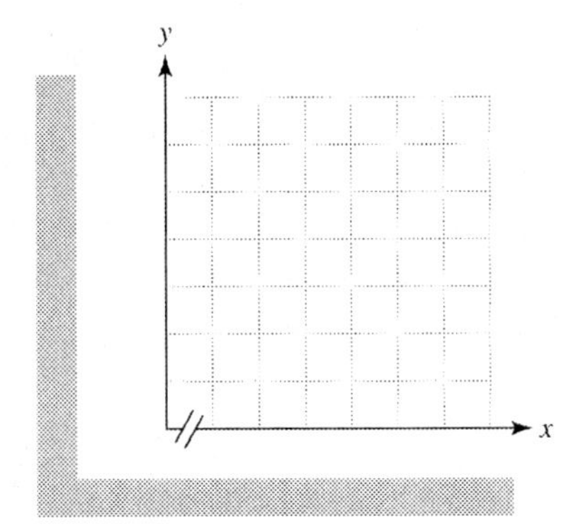

6. Use the data in the table to make a line graph. **6.** ________________

x	−4	−2	0	2	4
y	5	3	−4	−2	1

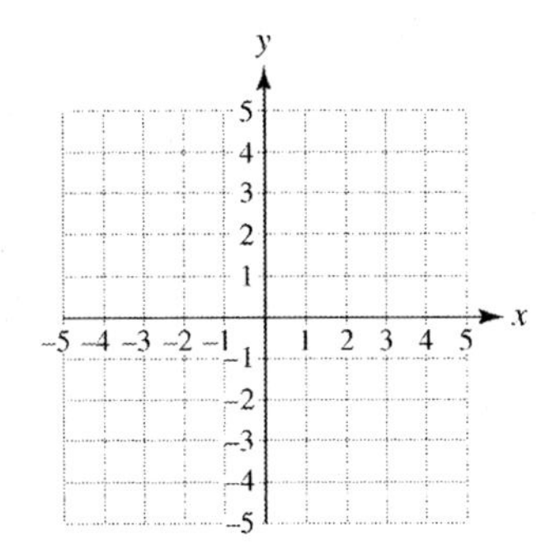

7. The line graph in the figure shows the per capita sugar consumption in the United States from 1950 to 2000.

7. ____________

Sugar Consumption in the United States

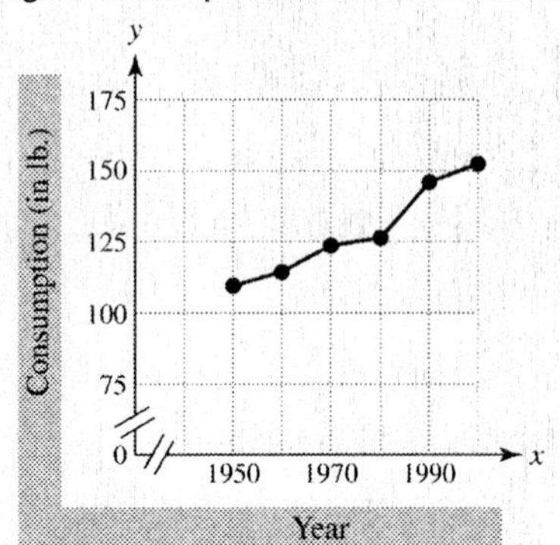

(a) Did sugar consumption ever decrease during this time period?

7. (a)____________

(b) Estimate the sugar consumption in 1960 and in 2000.

(b)____________

(c) Estimate the percent change in sugar consumption from 1960 to 2000.

(c)____________

Name: Course/Section: Instructor:

Chapter 3 Graphing Equations
3.2 Linear Equations

Basic Concepts ~ Table of Solutions ~ Graphing Linear Equations in Two Variables

STUDY PLAN

Read: Read Section 3.2 on pages 168-175 in your textbook or eText.

Practice: Do your assigned exercises in your ☐ Book ☐ MyMathLab ☐ Worksheets

Review: Keep your corrected assignments in an organized notebook and use them to review for the test.

Key Terms

Exercises 1-5: Use the vocabulary terms listed below to complete each statement. Note that some terms or expressions may not be used.

one	**no**
solution	**ordered pair**
standard form	**infinitely many**
linear equation in two variables	**table**

1. The graph of a(n) __________________ is a line.

2. Equations in two variables often have __________________ solution(s).

3. A(n) __________________ can be used to list solutions to an equation in two variables.

4. An ordered pair (x, y) whose x- and y- values satisfy the equation is called a(n) __________________ to an equation in two variables.

5. The __________________ of a linear equation in two variables is $Ax + By = C$, where A, B, and C are fixed numbers (constants) and A and B are not both equal to 0.

Basic Concepts

Exercises 1-3: Refer to Example 1 on page 169 in your text and the Section 3.2 lecture video.

Determine whether the given ordered pair is a solution to the given equation.

1. $y = x - 4,\ (-3,1)$ **1.** ________

2. $2x + y = 4;\ \left(-\frac{1}{2}, 3\right)$ **2.** ________

3. $-3x + 4y = 11;\ (-1,2)$ **3.** ________

Tables of Solutions

Exercises 4-6: Refer to Examples 2-4 on pages 170-171 in your text and the Section 3.2 lecture video.

4. Complete the table for the equation $y = 3x + 4$. **4.**

x	−4	−2	0	2
y				

5. Use $y = 0$, 4, 8, and 12 to make a table of solutions to $4x + 3y = 12$. **5.** ________

x				
y	0	4	8	12

6. The Asian-American population P in millions t years after the year 2000 is estimated by $P = 0.4t + 11.2$.

(a) Complete the table.

t	2	4	6	8	10
P					

(b) Use the table to determine the Asian-American population in the year 2008. **6.(b)** ________

Graphing Linear Equations in Two Variables

Exercises 7-10: Refer to Examples 5-7 on pages 171-174 in your text and the Section 3.2 lecture video.

7. Make a table of values for the equation $y = -3x$, and then use the table to graph this equation.

7. ________________

x	-2	-1	0	1
y				

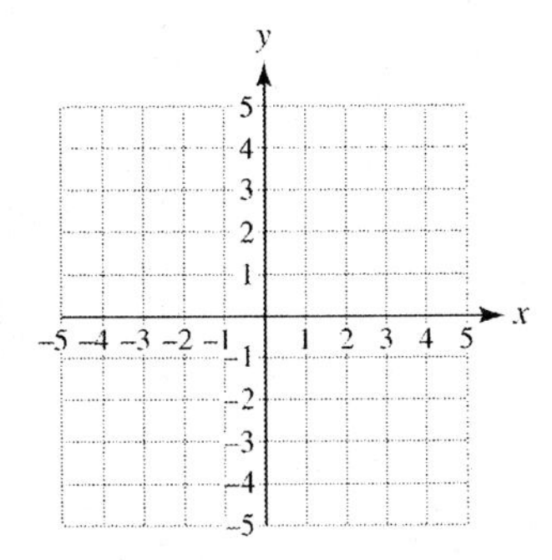

Graph each linear equation.

8. $y = -\frac{1}{2}x + 3$

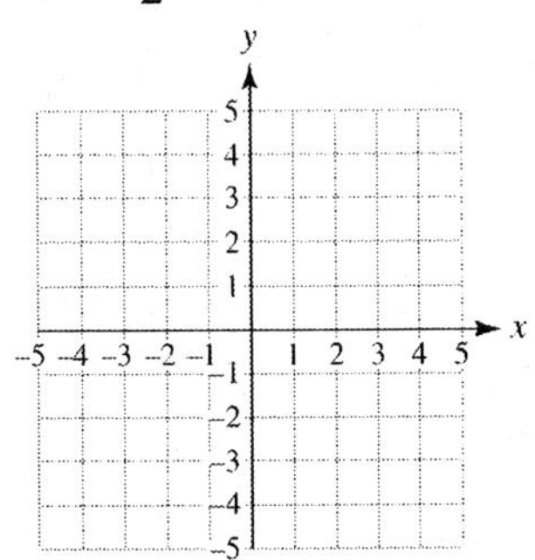

9. $x + y = -2$

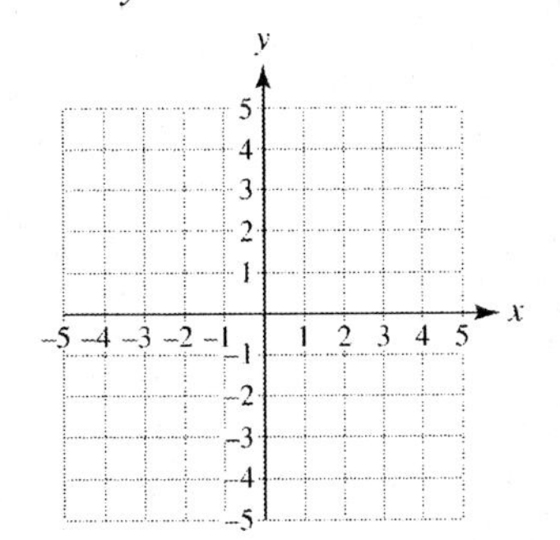

10. Graph the linear equation $3x - 5y = 15$ by solving for y first.

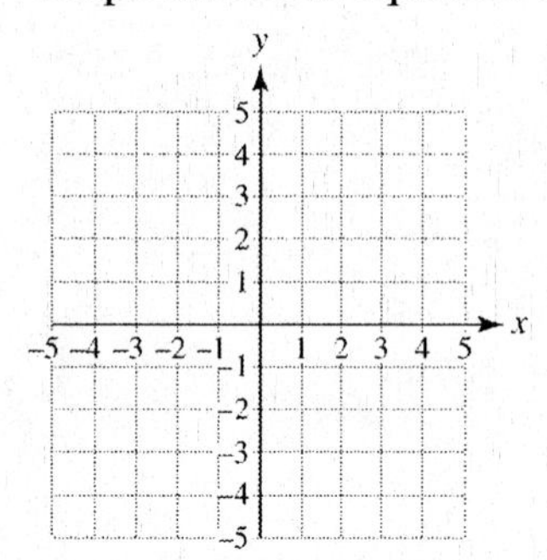

Name: Course/Section: Instructor:

Chapter 3 Graphing Equations
3.3 More Graphing of Lines

Finding Intercepts ~ Horizontal Lines ~ Vertical Lines

STUDY PLAN

Read: Read Section 3.3 on pages 178-186 in your textbook or eText.

Practice: Do your assigned exercises in your ☐ Book ☐ MyMathLab ☐ Worksheets

Review: Keep your corrected assignments in an organized notebook and use them to review for the test.

Key Terms

Exercises 1-4: Use the vocabulary terms listed below to complete each statement. Note that some terms or expressions may not be used.

x	y
$x = 0$	$y = 0$
x-axis	y-axis
x-coordinate	y-coordinate
vertical line	horizontal line

1. The equation of a ________________ with y-intercept b is $y = b$.

2. The ________________ of a point where a graph intersects the ________________ is called a y-intercept. To find a y-intercept, let ________________ in the equation and solve for ________________.

3. The equation of a ________________ with x-intercept k is $x = k$.

4. The ________________ of a point where a graph intersects the ________________ is called an x-intercept. To find the x-intercept, let ________________ in the equation and solve for ________________.

Finding Intercepts

Exercises 1-3: Refer to Examples 1-3 on pages 180-181 in your text and the Section 3.3 lecture video.

1. Use intercepts to graph $-5x+2y=10$.

1. 

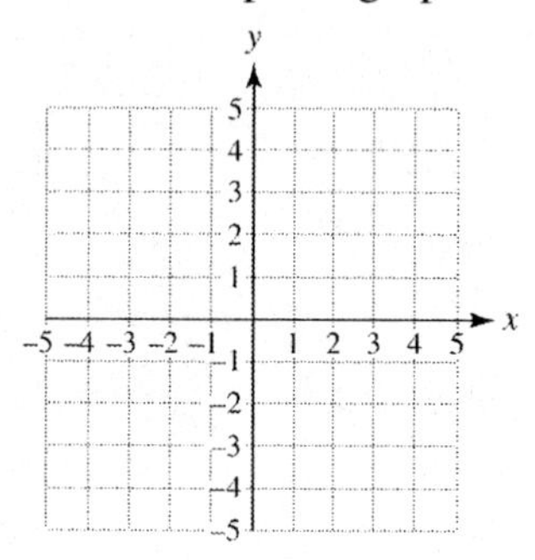

2. Complete the table for the equation $x-y=-2$. Then determine the x-intercept and y-intercept for the graph of the equation $x-y=-2$.

2. ______________

____

x	−2	−1	0	1	2
y					

3. A ball is thrown into the air. Its velocity v in feet per second after t seconds is given by $v=80-32t$. Assume $t \geq 0$ and $t \leq 2.5$.

(a) Graph the equation by finding the intercepts. Let t (time) correspond to the horizontal axis (x-axis) and v (velocity) correspond to the vertical axis (y-axis).

3. (a)

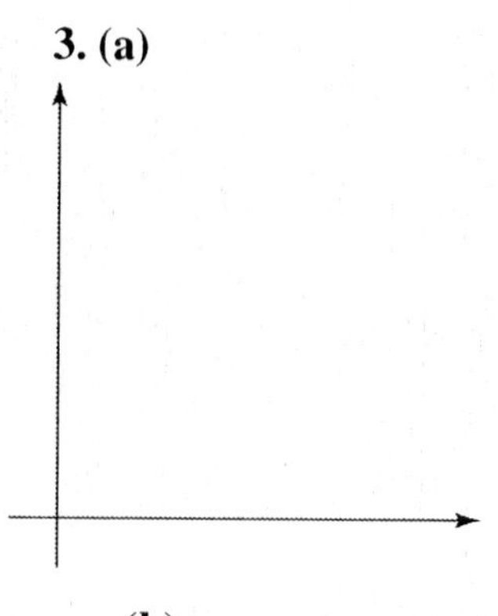

(b) Interpret each intercept.

(b) ______________

Horizontal Lines

Exercise 4: Refer to Example 4 on page 182 in your text and the Section 3.3 lecture video.

4. Graph the equation $y = -3$ and identify its y-intercept. **4.** ________________

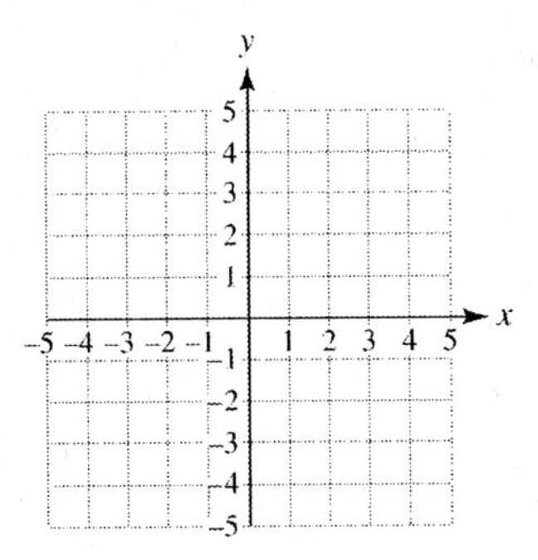

Vertical Lines

Exercise 5-10: Refer to Examples 5-7 on page 184-185 in your text and the Section 3.3 lecture video.

5. Graph the equation $x = -1$ and identify its x-intercept. **5.** ________________

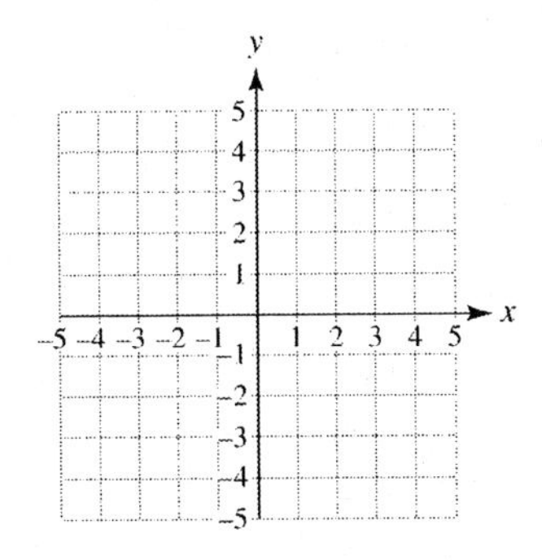

Write the equation of the line shown in each graph.

6. **6.** ________________

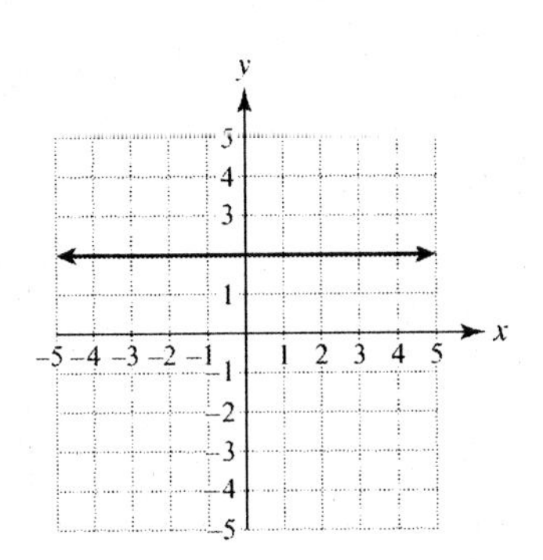

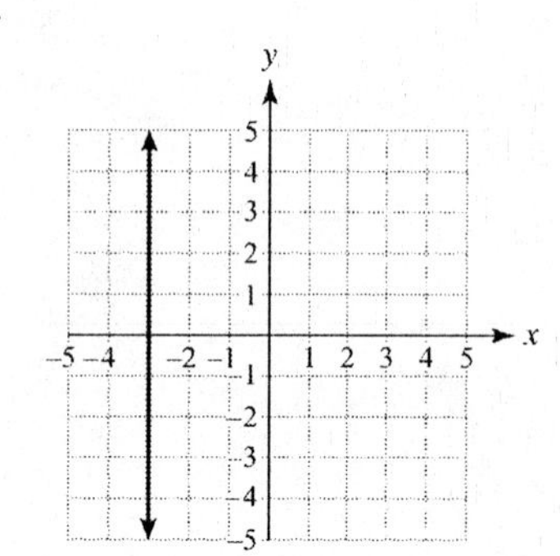

7. ______________

Find an equation for a line satisfying the given conditions.

8. Vertical, passing through $(-4,4)$

8. ______________

9. Horizontal, passing through $(5,1)$

9. ______________

10. Perpendicular to $y=-2$, passing through $(0,-6)$

10. ______________

Name: ______ **Course/Section:** ______ **Instructor:** ______

Chapter 3 Graphing Equations
3.4 Slope and Rates of Change

Finding Slopes of Lines ~ Slope as a Rate of Change

STUDY PLAN

Read: Read Section 3.4 on pages 190-198 in your textbook or eText.

Practice: Do your assigned exercises in your ☐ Book ☐ MyMathLab ☐ Worksheets

Review: Keep your corrected assignments in an organized notebook and use them to review for the test.

Key Terms

Exercises 1-8: Use the vocabulary terms listed below to complete each statement. Note that some terms or expressions may not be used.

rise	**run**
slope	**negative slope**
positive slope	**rate of change**
zero slope	**undefined slope**

1. A line with ________________ is horizontal.

2. If a line has ________________, it falls from left to right.

3. The ________________, or change in x, is $x_2 - x_1$.

4. The ________________ m of the line passing through the points (x_1, y_1) and (x_2, y_2) is $m = \frac{y_2 - y_1}{x_2 - x_1}$, where $x_1 \neq x_2$.

5. Slope measures the ________________ in a quantity.

6. The ________________, or change in y, is $y_2 - y_1$.

7. If a line has ________________, it rises from left to right.

8. A vertical line has ________________.

Finding Slopes of Lines

Exercises 1-9: Refer to Examples 1-6 on pages 191-194 in your text and the Section 3.4 lecture video.

1. Use the two points labeled in the figure to find the slope of the line. What are the rise and run between these two points? Interpret the slope in terms of rise and run.

1. 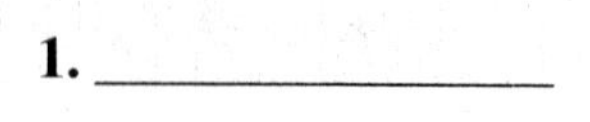

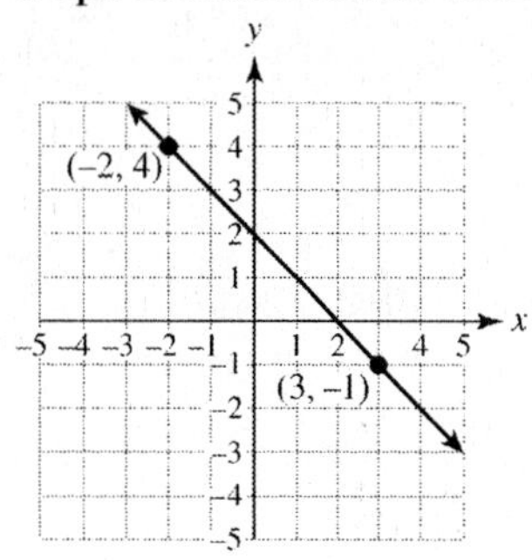

Calculate the slope of the line passing through each pair of points, if possible. Graph the line.

2. $(-4,1)$ and $(2,1)$

2. 

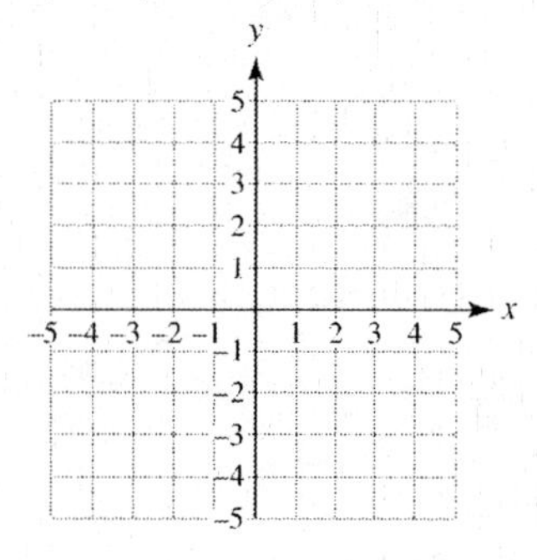

3. $(2,-2)$ and $(5,4)$

3. 

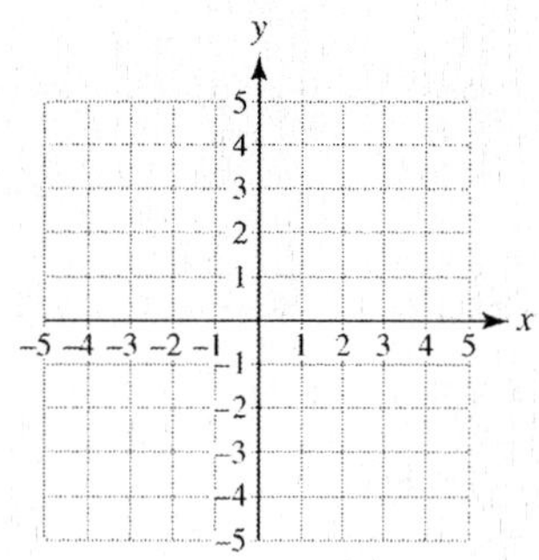

4. $(-3,4)$ and $(-3,-1)$ **4.** ______________

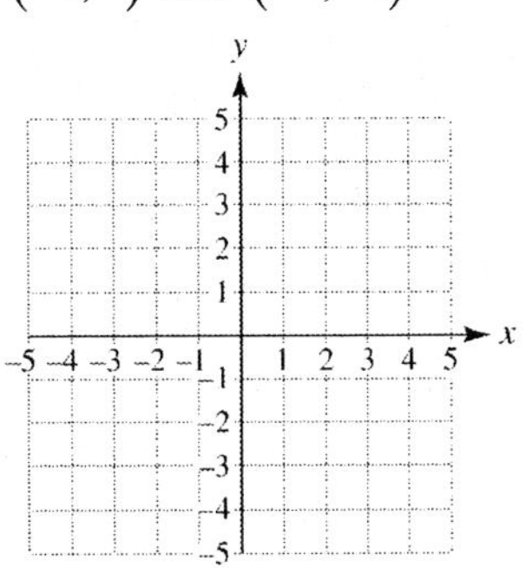

Find the slope of each line.

5. **5.** ______________

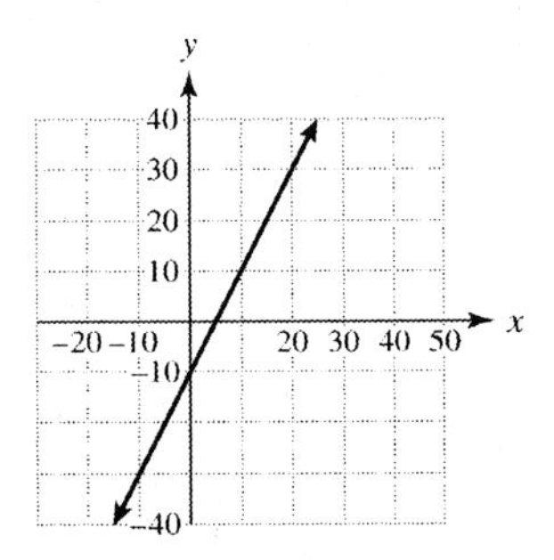

6. **6.** ______________

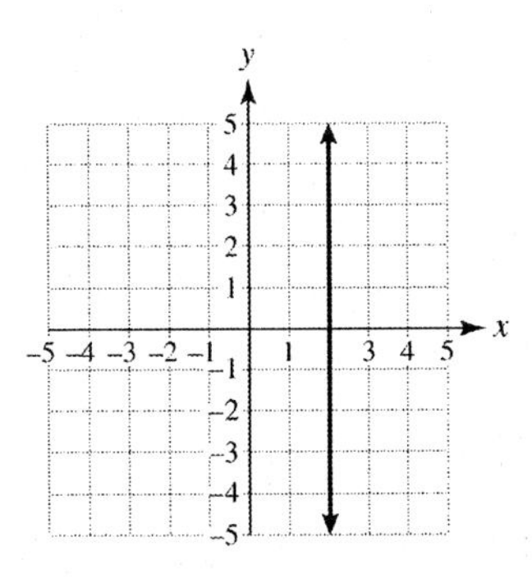

7. Sketch a line passing through the point $(-2,3)$ and having slope $-\frac{1}{4}$. **7.** ______________

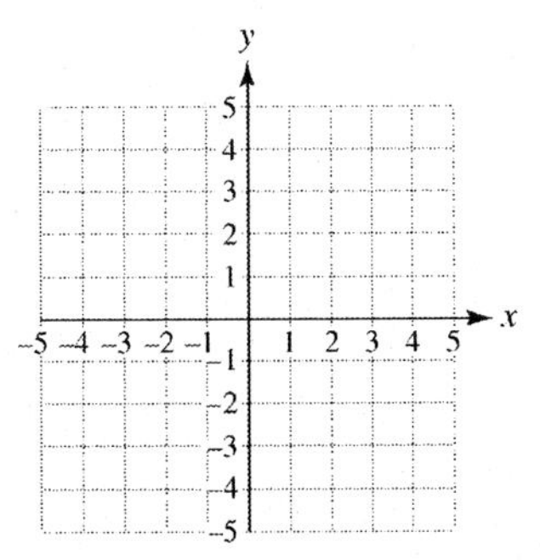

8. Sketch a line with slope -4 and y-intercept 1. **8.** ______________

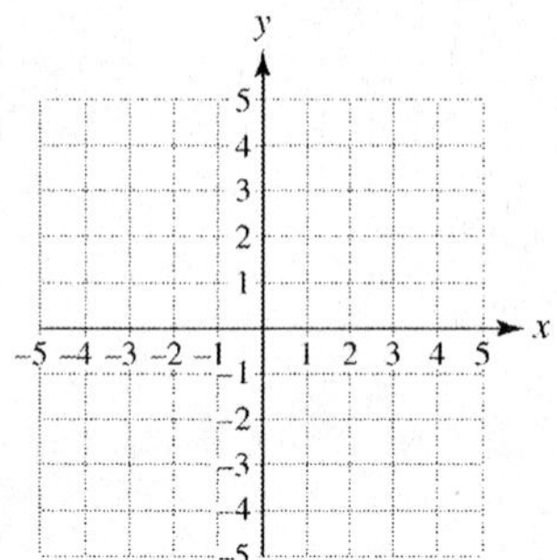

9. A line has slope -3 and passes through the first point listed in the table. Complete the table so that each point lies on the line. **9.** ______________

x	-2	-1	0	1
y	4			

Slope as a Rate of Change

Exercises 10-13: Refer to Examples 7-10 on pages 195-197 in your text and the Section 3.4 lecture video.

10. The distance y in miles that a cyclist training for a century ride is from home after x hours is shown in the figure.

Distance from Home

y
120
80
40
Distance (miles)
0 1 2 3 4 5 6
x
Time (hours)

(a) Find the y-intercept. What does it represent? **10. (a)**______________

(b) The graph passes through the point $(2,48)$. Discuss the meaning of this point. **(b)**______________

(c) Find the slope of this line. Interpret the slope as a rate of change. **(c)**______________

11. When a company manufactures 1000 game consoles, its profit is \$20,000, and when it manufactures 1500 game consoles, its profit is \$35,000.

(a) Find the slope of the line passing through $(1000, 20{,}000)$ and $(1500, 35{,}000)$.

11. (a)____________

(b) Interpret the slope as a rate of change.

(b)____________

12. The table lists the number of Baccalaureate degrees awarded after a private two-year college became a four-year institution.

Year	2002	2003	2005	2007	2009	2011
Baccalaureate Degrees	84	706	2206	2443	2726	3057

(a) Make a line graph of the data.

12. (a)____________

(b) Find the slope of each line segment.

(b)____________

(c) Interpret each slope as a rate of change.

(c)____________

13. During a storm, snow falls at the constant rates of 3 inches per hour from 2 P.M. to 5 P.M., 2 inches per hour from 5 P.M. to 7 P.M. and $\frac{1}{2}$ per hour from 7 P.M. to 9 P.M.

(a) Sketch a graph that shows the total accumulation of snowfall from 2 P.M. to 9 P.M.

13. (a)____________

(b) What does the slope of each line segment represent?

(b)____________

Name: **Course/Section:** **Instructor:**

Chapter 3 Graphing Equations
3.5 Slope-Intercept Form

Basic Concepts ~ Finding Slope-Intercept Form ~ Parallel and Perpendicular Lines

STUDY PLAN

Read: Read Section 3.5 on pages 204-211 in your textbook or eText.

Practice: Do your assigned exercises in your ☐ Book ☐ MyMathLab ☐ Worksheets

Review: Keep your corrected assignments in an organized notebook and use them to review for the test.

Key Terms

Exercises 1-7: Use the vocabulary terms listed below to complete each statement. Note that some terms or expressions may not be used.

parallel	**slope**
point-slope	**zero slope**
perpendicular	**slope-intercept**
undefined slope	**negative reciprocal**

1. A line with __________________ is horizontal.

2. Two nonvertical __________________ lines have the same slope.

3. The __________________ m of the line passing through the points (x_1, y_1) and (x_2, y_2) is $m = \frac{y_2 - y_1}{x_2 - x_1}$, where $x_1 \neq x_2$.

4. If two lines have slopes m_1 and m_2 such that $m_1 \cdot m_2 = -1$, then they are __________________ lines.

5. The __________________ form of a line with slope m and y-intercept b is given by $y = mx + b$.

6. The slopes of two perpendicular lines are __________________(s) of each other.

7. A vertical line has __________________.

Finding Slope-Intercept Form

Exercises 1-7: Refer to Examples 1-5 on pages 205-208 in your text and the Section 3.5 lecture video.

For each graph write the slope-intercept form of the line.

1.   **1.** ______________

2.   **2.** ______________

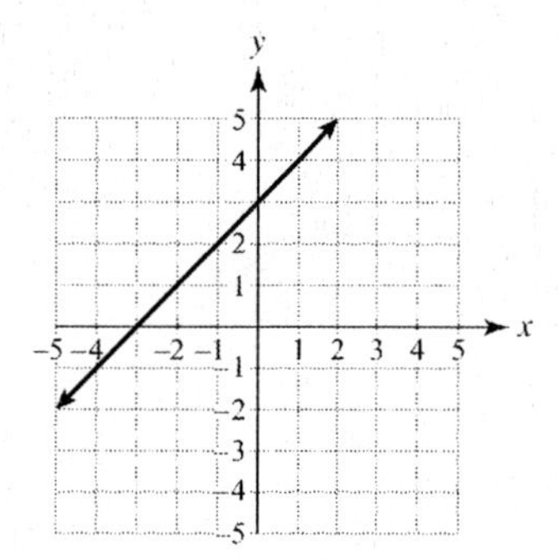

3. Sketch a line with slope $\frac{1}{3}$ and y-intercept –4. Write its slope-intercept form.   **3.** ______________

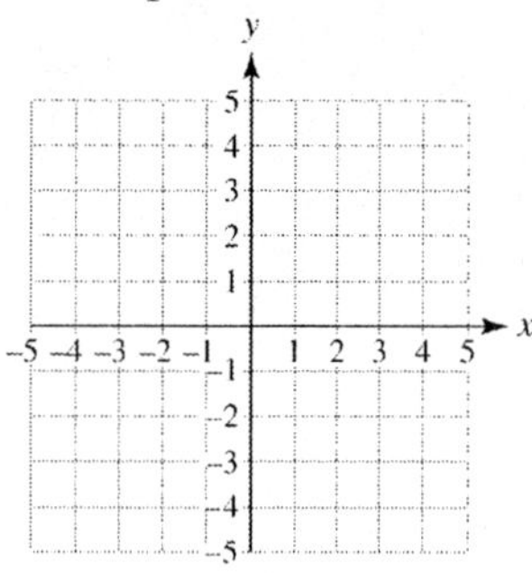

Write each equation in slope-intercept form. Then give the slope and y-intercept of the line.

4. $3y - 5x = 15$

4. ______________
____
____

5. $x = -3y + 6$

5. ______________
____
____

6. Write the equation $3x - y = 2$ in slope-intercept form and then graph it.

6. ______________

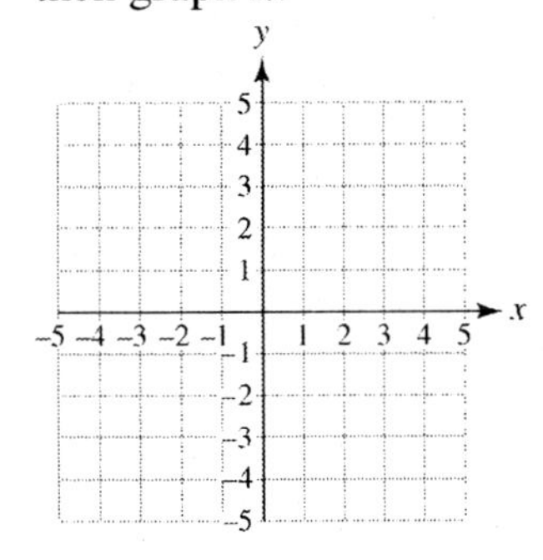

7. Production of a certain item involves fixed costs of \$34,000 plus \$120 for each item made.

(a) How much does it cost to produce 2000 items?

7. (a)______________

(b) Write the slope-intercept form that gives the cost to produce x items.

(b)______________

(c) If the cost is \$454,000, how many items were produced?

(c)______________

Parallel and Perpendicular Lines

Exercises 8-12: Refer to Examples 6-8 on pages 208-210 in your text and the Section 3.5 lecture video.

8. Find the slope-intercept form of a line parallel to $y = 2x - 7$ and passing through the point $(-2,1)$. Sketch each line in the same xy-plane.

8. 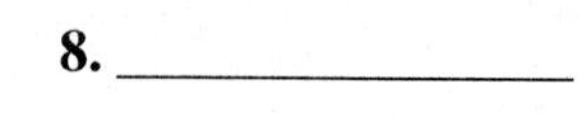

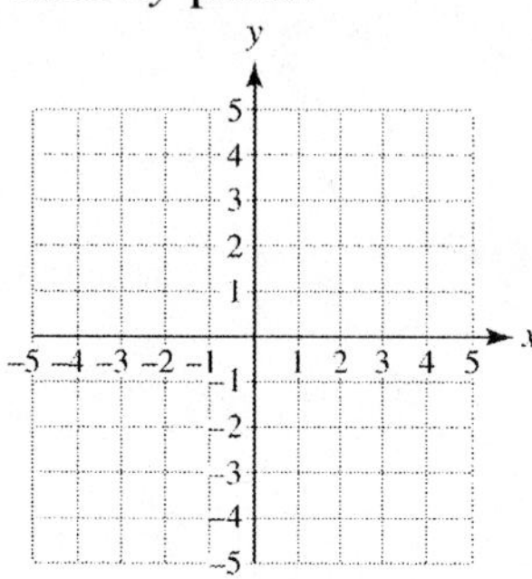

For each of the given lines, find the slope-intercept form of a line passing through the origin that is perpendicular to the given line.

9. $y = -4x$

9. ______________

10. $y = \frac{2}{3}x - 4$

10. ______________

11. $5x + 2y = -10$

11. ______________

12. Find the slope-intercept form of a line perpendicular to $y = \frac{3}{4}x + 2$ and passing through the point $(3,-1)$. Sketch each line in the same xy-plane.

12. 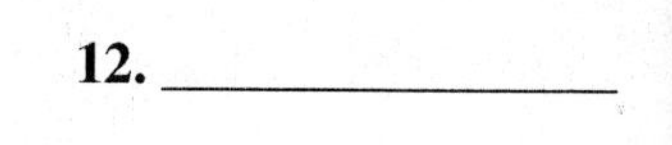

Name: Course/Section: Instructor:

Chapter 3 Graphing Equations
3.6 Point-Slope Form

Derivation of Point-Slope Form ~ Finding Point-Slope Form ~ Applications

STUDY PLAN

Read: Read Section 3.6 on pages 214-221 in your textbook or eText.

Practice: Do your assigned exercises in your ☐ Book ☐ MyMathLab ☐ Worksheets

Review: Keep your corrected assignments in an organized notebook and use them to review for the test.

Key Terms

Exercises 1-3: Use the vocabulary terms listed below to complete each statement. Note that some terms or expressions may not be used.

slope
point-slope
slope-intercept

1. The line with slope m passing through the point (x_1, y_1) is given by $y - y_1 = m(x - x_1)$, or equivalently, $y = m(x - x_1) + y_1$. This is called the ________________ form of a line.

2. The ________________ m of the line passing through the points (x_1, y_1) and (x_2, y_2) is $m = \frac{y_2 - y_1}{x_2 - x_1}$, where $x_1 \neq x_2$.

3. The ________________ form of a line with slope m and y-intercept b is given by $y = mx + b$.

Derivation of Point-Slope Form

Writing Exercise: Use the slope formula to derive the point-slope form of a line.

Finding Point-Slope Form

Exercises 1-8: Refer to Examples 1-5 on pages 215-219 in your text and the Section 3.6 lecture video.

Use the labeled point in each figure to write a point-slope form for the line and then simplify it to the slope-intercept form.

1.

1. 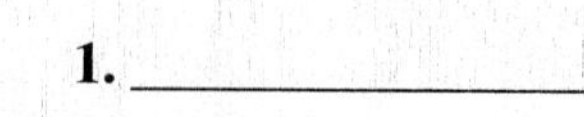

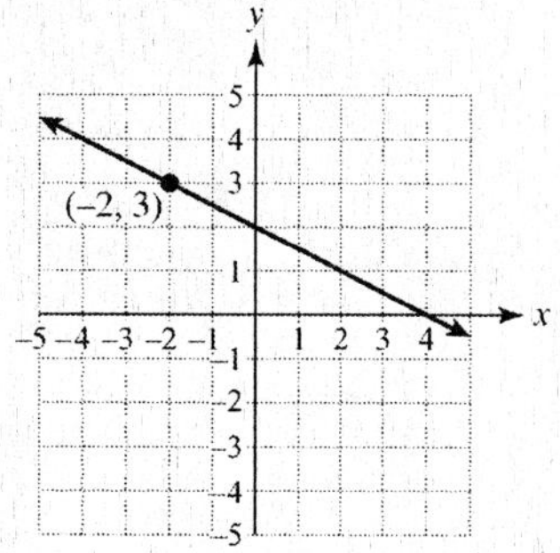

2.

2. 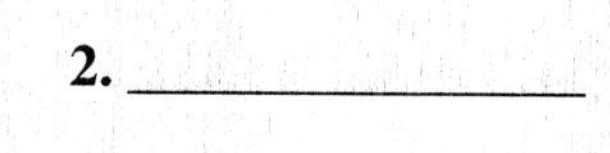

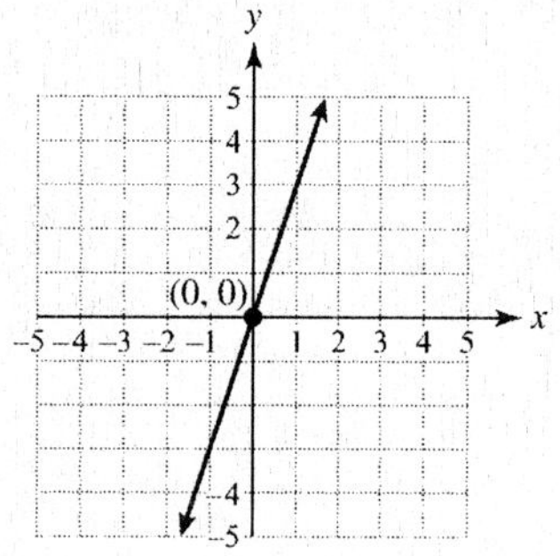

3. Find a point-slope form for a line passing through the point $(2,-2)$ with slope $-\frac{1}{4}$. Does the point $(-2,3)$ lie on this line?

3. ______________

____

4. Use the point-slope form to find an equation of the line passing through the points $(2,-4)$ and $(1,1)$

4. ______________

Find the slope-intercept form for the line that satisfies the given conditions.

5. Slope $-\frac{1}{2}$, passing through $(4,-1)$ **5.** ______________

6. x-intercept -4, y-intercept 1 **6.** ______________

7. Perpendicular to $3x+4y=5$, passing through $\left(\frac{3}{4},-2\right)$ **7.** ______________

8. The points in the table lie on a line. Find the slope-intercept form of the line. **8.** ______________

x	-2	-1	0	1
y	7	4	1	-2

Applications

Exercises 9-10: Refer to Examples 6-7 on pages 219-220 in your text and the Section 3.6 lecture video.

9. In 2005, a self-employed man earned \$56,000. In 2011, he earned \$74,000.

(a) Find a point-slope form of the line passing through the points $(2006, 56)$ and $(2011, 74)$. **9. (a)**______________

(b) Interpret the slope as a rate of change. **(b)**______________

(c) Estimate the man's salary in the year 2014. **(c)**______________

10. A tank is being emptied by a pump that removes water at a constant rate. After 2 hours, the tank contains 4500 gallons of water, and after 5 hours, the tank contains 2250 gallons of water.

(a) How fast is the pump removing water? **10.(a)**____________

(b) Find the slope-intercept form of a line that models the amount of water in the tank. Interpret the slope. **(b)**____________

(c) Find the y-intercept and the x-intercept. Interpret each. **(c)**____________

(d) Sketch a graph of the amount of water in the tank during the first 8 hours. **(d)**

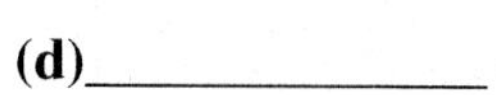

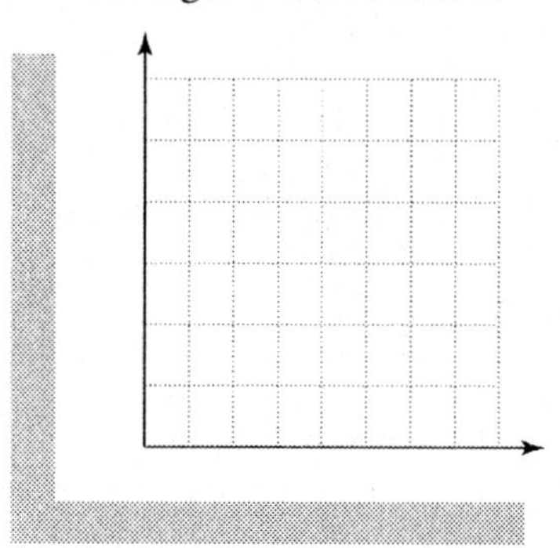

(e) The point $(4, 3000)$ lies on the graph. Explain its meaning. **(e)**____________

Name: **Course/Section:** **Instructor:**

Chapter 3 Graphing Equations
3.7 Introduction to Modeling

Basic Concepts ~ Modeling Linear Data

STUDY PLAN

Read: Read Section 3.7 on pages 225-229 in your textbook or eText.

Practice: Do your assigned exercises in your ☐ Book ☐ MyMathLab ☐ Worksheets

Review: Keep your corrected assignments in an organized notebook and use them to review for the test.

Basic Concepts

Exercise 1: Refer to Example 1 on page 226 in your text and the Section 3.7 lecture video.

1. After driving 100 miles, a traveler begins to track his mileage. The table below shows the cumulative distance driven after selected days. Does the equation $D = 85x + 100$ model the data exactly? Explain.

1. ____________

x	3	6	7
D	355	615	695

Modeling Linear Data

Exercises 2-7: Refer to Examples 2-5 on pages 227-229 in your text and the Section 3.7 lecture video.

2. The table shows the number of miles y traveled by a truck on x gallons of gasoline.

x	3	6	9	12
y	36	72	108	144

(a) Plot the data in the xy-plane. Be sure to label each axis.

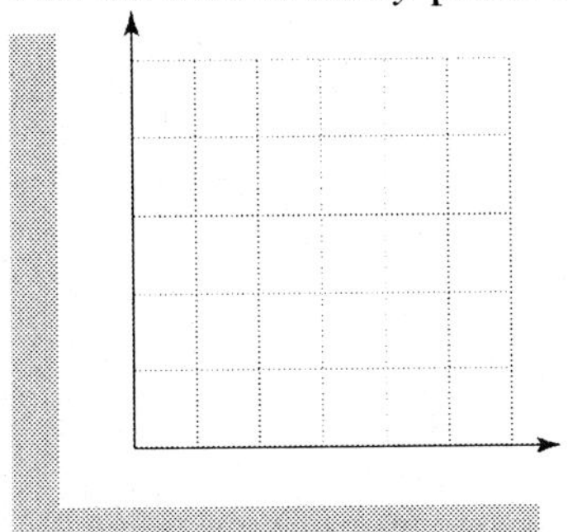

(b) Sketch a line that models the data.

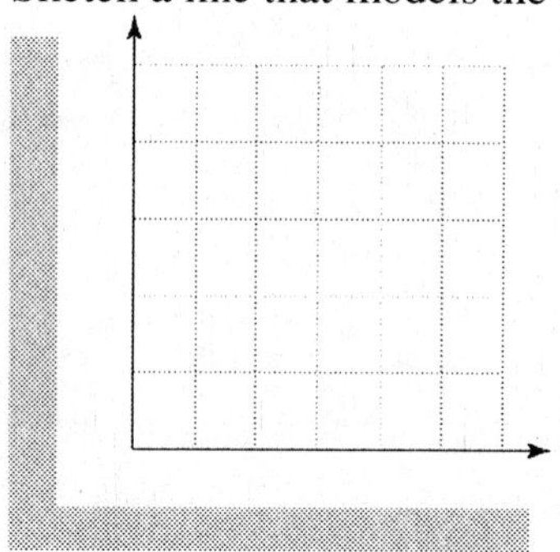

(c) Find the equation of the line and interpret the slope of the line. **2.(c)**________________

(d) How far could this truck travel on 16 gallons of gasoline? **(d)**________________

3. The table contains ordered pairs that can be modeled approximately by a line.

x	-3	-1	0	1	3
y	2	1	0	-1	-2

(a) Plot the data. Could a line pass through all five points? **3.(a)**________________

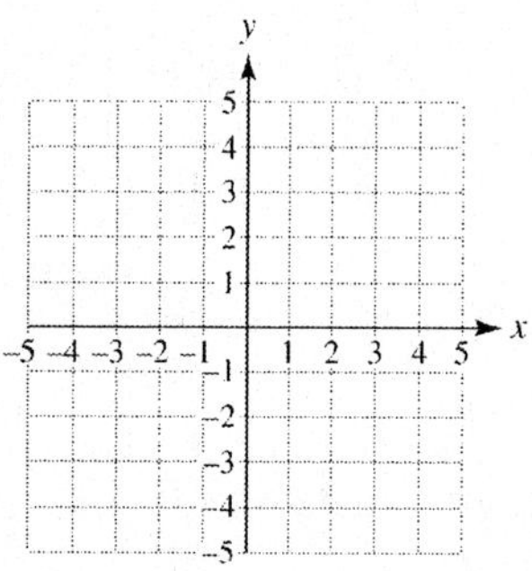

(b) Sketch a line that models the data and then determine its equation. **(b)**________________

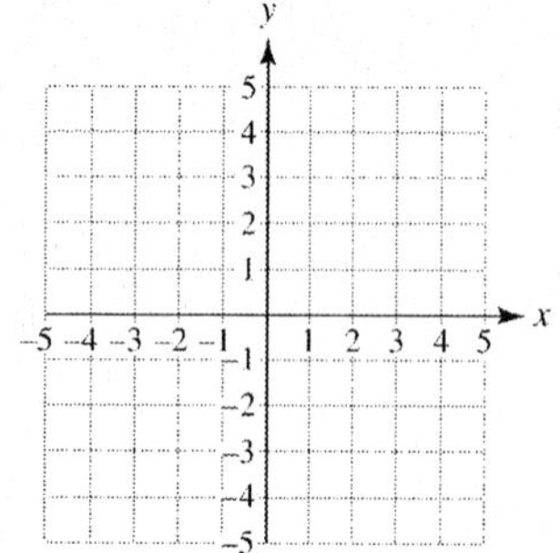

4. According to data from the 2011 National Diabetes Fact Sheet, at the beginning of 2011, a total of 18.8 million people had been diagnosed with diabetes, and the disease was growing at a rate of 1.9 million new diagnoses each year.

(a) Write a linear equation $N = mx + b$ that models the total number. of people N in millions that have been diagnosed with diabetes x years after January 1, 2011, **4.(a)**________________

(b) Estimate N at the beginning of 2015. **(b)**________________

Find a linear equation in the form $y = mx + b$ ***that models the quantity y after x days.***

5. A quantity y is initially 2500 and remains constant. **5.** ________________

6. A quantity y is initially 700 and increases at a rate of 28 per day. **6.** ________________

7. A quantity y is initially 480 and decreases at a rate of 10 per day. **7.** ________________

Name: **Course/Section:** **Instructor:**

Chapter 4 Systems of Linear Equations in Two Variables
4.1 Solving Systems of Linear Equations Graphically and Numerically

Basic Concepts ~ Solutions to Systems of Equations

STUDY PLAN

Read: Read Section 4.1 on pages 246-254 in your textbook or eText.

Practice: Do your assigned exercises in your ☐ Book ☐ MyMathLab ☐ Worksheets

Review: Keep your corrected assignments in an organized notebook and use them to review for the test.

Key Terms

Exercises 1-6: Use the vocabulary terms listed below to complete each statement. Note that some terms or expressions may not be used. Some terms may be used more than once.

parallel	**inconsistent**
identical	**independent**
consistent	**solution to a system**
dependent	**intersection-of-graphs**
intersecting	**system of linear equations**

1. A system of equations with exactly one solution is a(n) ________________ system with ________________ equations. Graphing the system results in ________________lines.

2. The graphical technique of solving two equations is sometimes called the ________________ method.

3. A system of equations with no solutions is a(n) ________________ system. Graphing the system results in ________________ lines.

4. A(n) ________________ of two equations is an ordered pair (x, y) that makes both equations true.

5. A(n) ________________ in two variables is written such that each equation is a linear equation in two variables.

6. A system of equations with infinitely many solutions is a(n) ________________ system with ________________ equations. Graphing the system results in ________________ lines.

Basic Concepts

Exercises 1-3: Refer to Examples 1-2 on pages 247-248 in your text and the Section 4.1 lecture video.

1. The equation $P = 15x$ calculates an employee's pay for working x hours at \$15 per hour. Use the intersection-of-graphs method to find the number of hours that the employee worked if the amount paid is \$75.

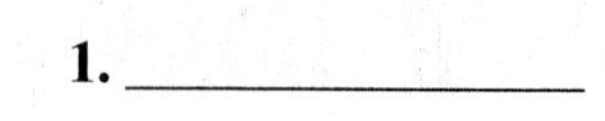

1. ________________

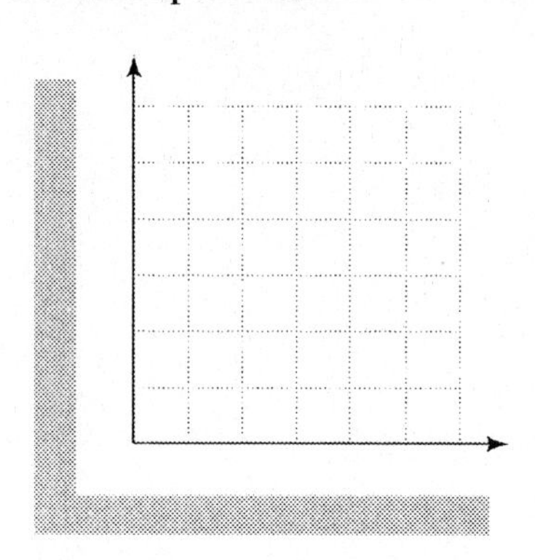

Use a graph to find the x-value when $y = -4$.

2. $y = -x - 2$

2. ________________

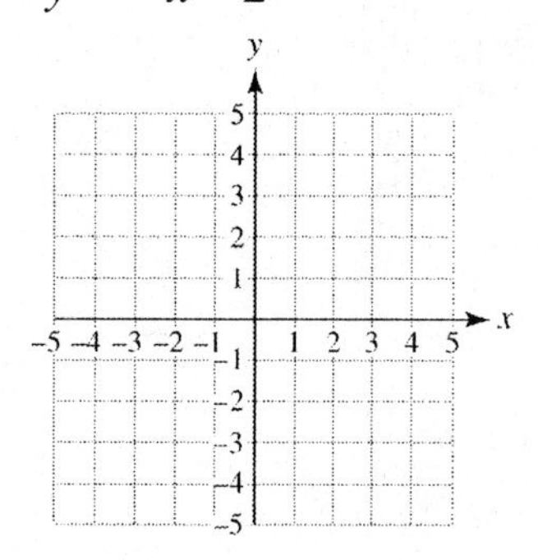

3. $4x - 3y = 12$

3. ________________

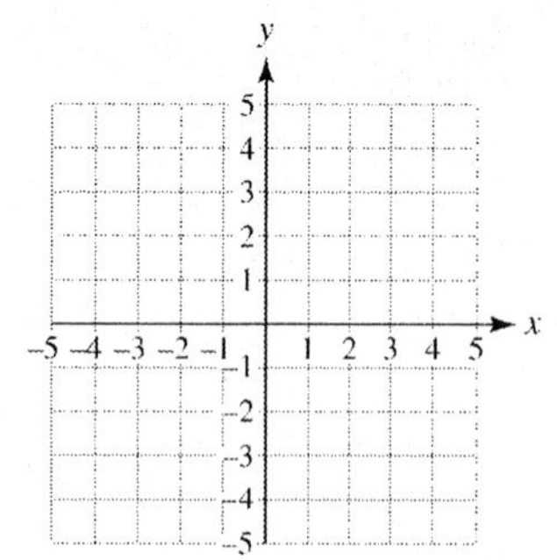

Solutions to Systems of Equations

Exercises 4-10: Refer to Examples 3-7 on pages 250-252 in your text and the Section 4.1 lecture video.

Graphs of two equations are shown. State the number of solutions to each system of equations. Then state whether the system is consistent or inconsistent. If it is consistent, state whether the equations are dependent or independent.

4. **4.** ______________

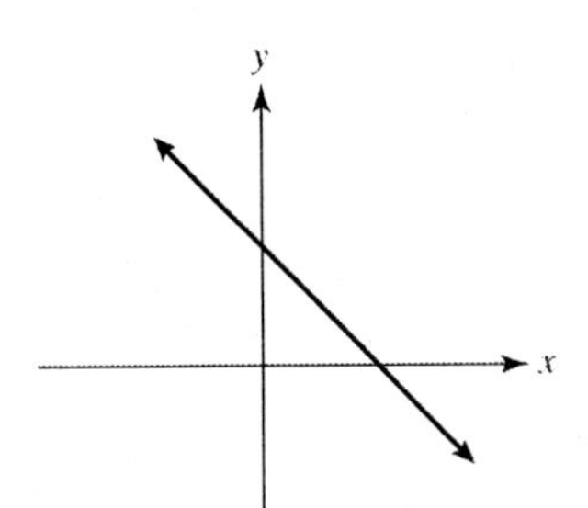

5. **5.** ______________

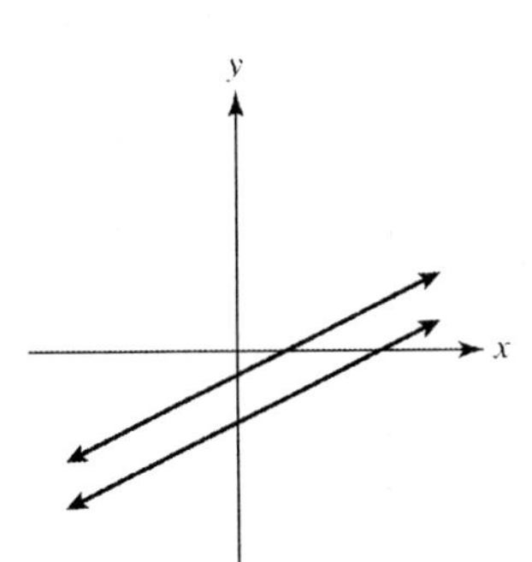

6. **6.** ______________

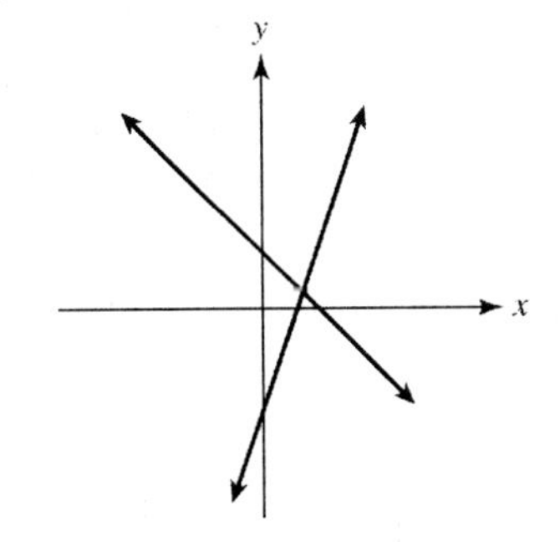

7. Determine whether $(-5,-3)$ or $(-2,-2)$ is the solution to the system of equations

$$x-2y=1$$
$$x-3y=4.$$

7. ____________

8. Solve the system of linear equations

$$x-y=-3$$
$$3x-y=-3$$

with a graph and with a table of values.

8. ____________

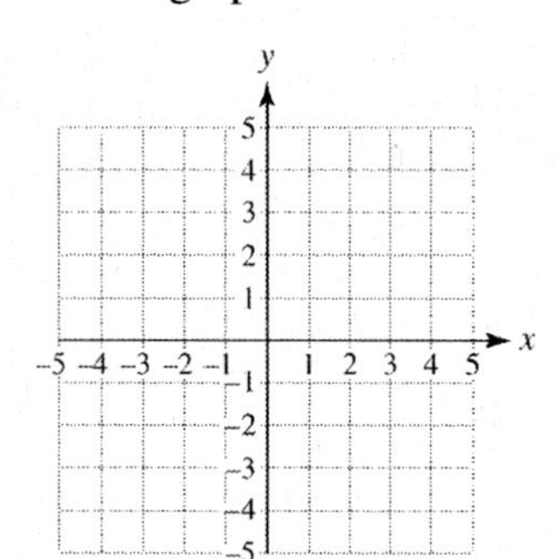

x					
$y=$					
$y=$					

9. Solve the system of equations graphically.

$$y=-2x$$
$$3x-y=5$$

9. ____________

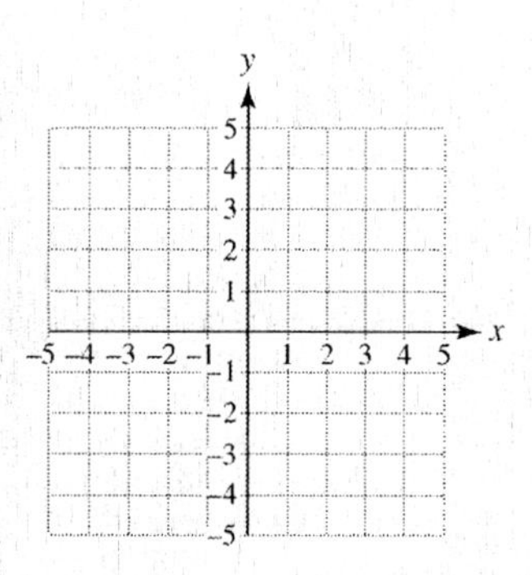

10. Last year, there were 12,000 students at a local college. There were 2000 more female students than male students. How many male and female students attended the college?

10. ____________

____

Name: **Course/Section:**   **Instructor:**

Chapter 4 Systems of Linear Equations in Two Variables
4.2 Solving Systems of Linear Equations by Substitution

The Method of Substitution ~ Recognizing Other Types of Systems ~ Applications

STUDY PLAN

Read: Read Section 4.2 on pages 257-262 in your textbook or eText.

Practice: Do your assigned exercises in your ☐ Book ☐ MyMathLab ☐ Worksheets

Review: Keep your corrected assignments in an organized notebook and use them to review for the test.

Key Terms

Exercises 1-3: Use the vocabulary terms listed below to complete each statement. Note that some terms or expressions may not be used.

parallel
identical
no solutions
method of substitution
infinitely many solutions

1. The technique of substituting an expression for a variable and solving the resulting equation is called the ___________________.

2. If the process of solving a system of equations results in a statement that is always false, like $7 = 4$, we conclude that the system has ___________________. Graphing the system results in ___________________ lines.

3. If the process of solving a system of equations results in a statement that is always true, like $-3 = -3$, we conclude that the system has ___________________. Graphing the system results in ___________________ lines.

The Method of Substitution

Exercises 1-4: Refer to Examples 1-2 on pages 258-259 in your text and the Section 4.2 lecture video.

Solve each system of equations.

1. $4x - y = -4$

$y = 2x$

1. ________________

2. $x - 2y = 1$

$x = 4y$

2. ________________

3. $x + y = 5$

$x - 4y = 5$

3. ________________

4. $4x - y = 7$

$2x + 5y = 9$

4. ________________

Recognizing Other Types of Systems

Exercises 5-6: Refer to Example 3 on page 260 in your text and the Section 4.2 lecture video.

If possible, use substitution to solve the system of equations. Then use graphing to help explain the result.

5. $x - 3y = 1$

$2x - 6y = 2$

5. ________________

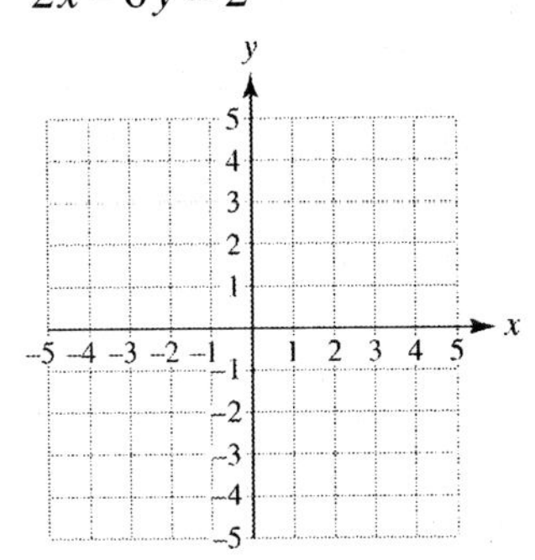

6. $2x - y = -2$

$-6x + 3y = 4$

6. ________________

Applications

Exercises 7-9: Refer to Examples 4-6 on pages 260-262 in your text and the Section 4.2 lecture video.

7. A company places a total of 450 advertisements using radio and television. The number of radio ads is twice the number of television ads. Find the number of radio ads and the number of television ads.

7. ________________

____

8. For a wedding reception, the hostess served desserts and hors d'oeuvres. A total of \$2500 was spent on food. \$700 more was spent on hors d'oeuvres than on desserts. How much was spent on hors d'oeuvres? How much was spent on desserts?

8. ______________

____

9. An airplane flies 1680 miles into (or against) the wind in 4 hours. The return trip takes $3\frac{1}{2}$ hours. Find the speed of the airplane with no wind and the speed of the wind.

9. ______________

____

Name: **Course/Section:** **Instructor:**

Chapter 4 Systems of Linear Equations in Two Variables
4.3 Solving Systems of Linear Equations by Elimination

The Elimination Method ~ Recognizing Other Types of Systems ~ Applications

STUDY PLAN

Read: Read Section 4.3 on pages 265-273 in your textbook or eText.

Practice: Do your assigned exercises in your ☐ Book ☐ MyMathLab ☐ Worksheets

Review: Keep your corrected assignments in an organized notebook and use them to review for the test.

Key Terms

Exercises 1-3: Use the vocabulary terms listed below to complete each statement. Note that some terms or expressions may not be used.

consistent
no solutions
dependent
elimination method
independent
inconsistent
infinitely many solutions

1. If the process of solving a system of equations results in a statement that is always false, like $0 = 5$, we conclude that the system has ________________ and is a(n) ________________ system.

2. If the process of solving a system of equations results in a statement that is always true, like $0 = 0$, we conclude that the system has ________________ and is a(n) ________________ system with ________________ equations.

3. The ________________ is based on the addition property of equality.

The Elimination Method

Exercises 1-6: Refer to Examples 1-4 on pages 266-270 in your text and the Section 4.3 lecture video.

Solve each system of equations. Check each solution.

1. $x - 2y = -4$
$x + 3y = 1$

1. ______________

2. $3a - 2b = 2$
$4a - 5b = -2$

2. ______________

3. $-r - 4t = -8$
$3r + t = -9$

3. ______________

4. $2x + 6y = 4$
$5x + 3y = 1$

4. ______________

5. Solve the system of equations two times, first by eliminating x and then by eliminating y.

$3y = -3 - 4x$
$2x = -5y - 19$

5. ______________

6. Solve the system of equations symbolically, graphically, and numerically.

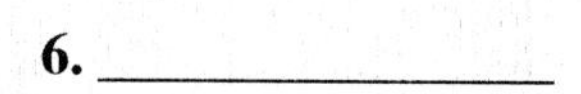

6. ______________

$x - y = 4$
$4x + y = 1$

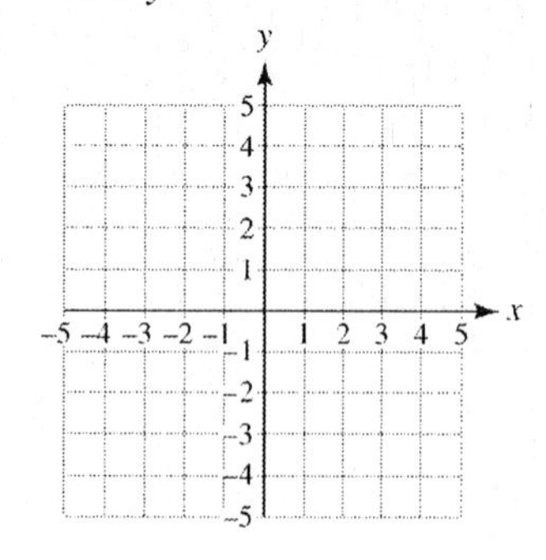

x	−2	−1	0	1	2
$y =$					
$y =$					

Recognizing Other Types of Systems

Exercises 7-8: Refer to Example 5 on pages 270-271 in your text and the Section 4.3 lecture video.

Solve each system of equations by using the elimination method. Then graph the system.

7. $x - 4y = 6$
$-x + 4y = -3$

7. ________________

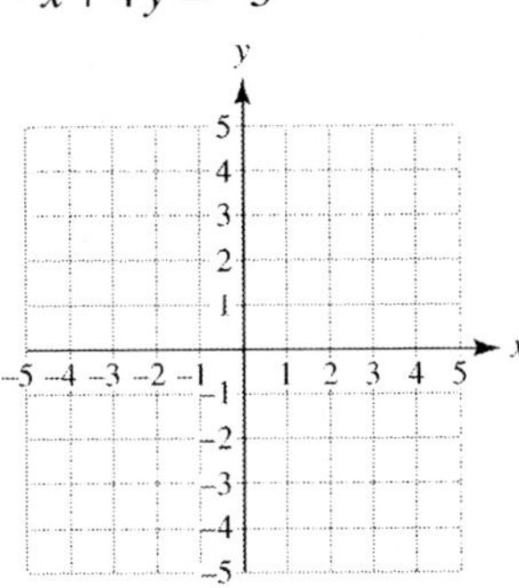

8. $x - 3y = 4$
$-2x + 6y = -8$

8. ________________

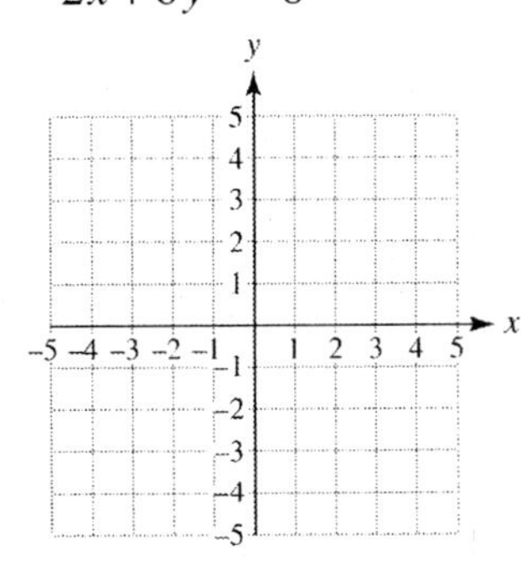

Applications

Exercises 9-10: Refer to Examples 6-7 on pages 271-272 in your text and the Section 4.3 lecture video.

9. Last year, enrollment at a local college increased by 1800 students. There were 400 more new female students than new male students.

(a) How many new students were women?

(b) How many new students were men?

9. (a)________________

(b)________________

10. During strenuous exercise, an athlete can burn 12 calories per minute on a treadmill and 8.5 calories per minute on an elliptical machine. If an athlete uses both machines and burns 325 calories in a 30-minute workout, how many minutes does the athlete spend on each machine?

10. ________________

____

Understanding Concepts through Multiple Approaches
(For additional practice, visit MyMathLab.)

11. Solve the system of equations.

$$4x + 2y = 4$$
$$2x - y = -6$$

(a) Solve algebraically.

(b) Solve numerically using the table shown.

x	−2	−1	0	1	2
$y =$					
$y =$					

(c) Solve visually.

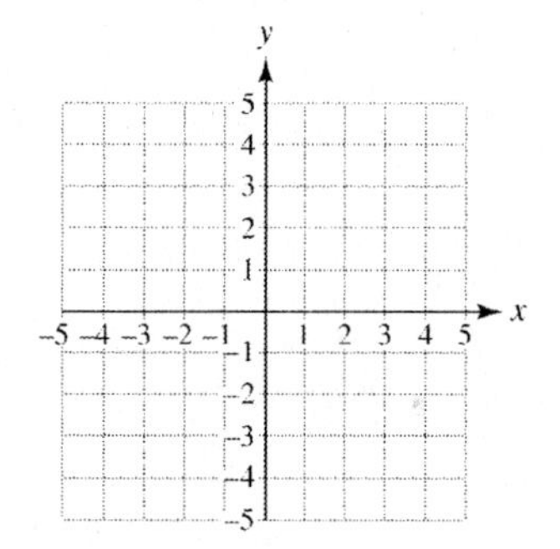

Did you get the same result using each method? Which method do you prefer? Explain why.

Name: ____________ **Course/Section:** ____________ **Instructor:** ____________

Chapter 4 Systems of Linear Equations in Two Variables
4.4 Systems of Linear Inequalities

Basic Concepts ~ Solutions to One Inequality ~ Solutions to Systems of Inequalities ~ Applications

STUDY PLAN

Read: Read Section 4.4 on pages 276-284 in your textbook or eText.

Practice: Do your assigned exercises in your ☐ Book ☐ MyMathLab ☐ Worksheets

Review: Keep your corrected assignments in an organized notebook and use them to review for the test.

Key Terms

Exercises 1-5: Use the vocabulary terms listed below to complete each statement. Note that some terms or expressions may not be used.

solution
test point
solution set
linear inequality
system of linear inequalities

1. A(n) __________________ may be used to determine in which region of the xy-coordinate plane solutions to a linear inequality lie.

2. When the equals sign in any linear equation is replaced with $<$, $>$, $\leq$, or $\geq$, a(n) __________________ in two variables results.

3. A(n) __________________ to a linear inequality in two variables is an ordered pair (x, y) that makes the inequality a true statement.

4. A(n) __________________ in two variables consists of two or more linear inequalities.

5. The __________________ is the set of all solutions to an inequality.

Solutions to One Inequality

Exercises 1-5: Refer to Examples 1-2 on pages 278-279 in your text and the Section 4.4 lecture video.

Write a linear inequality that describes each shaded region.

1.

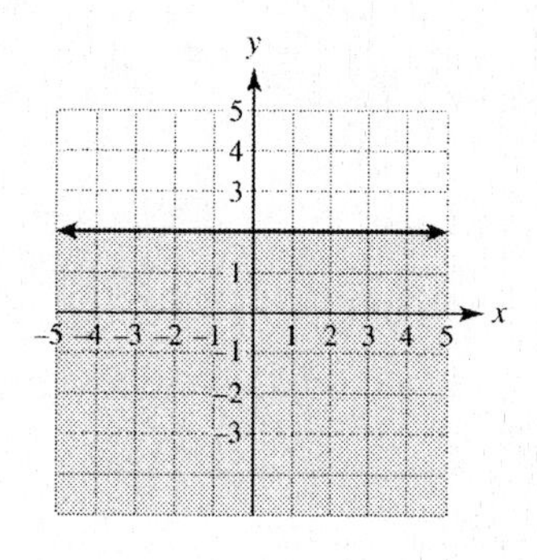

1. 

2.

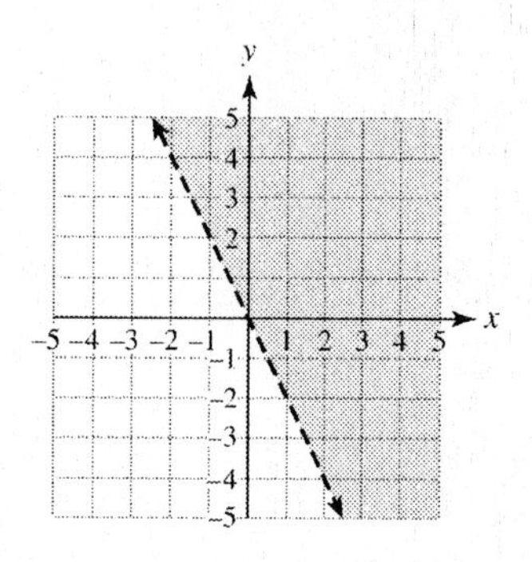

2. 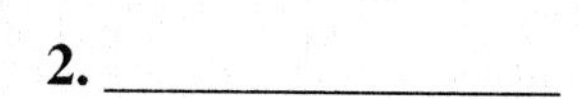

Shade the solution set for each inequality.

3. $x \leq 2$

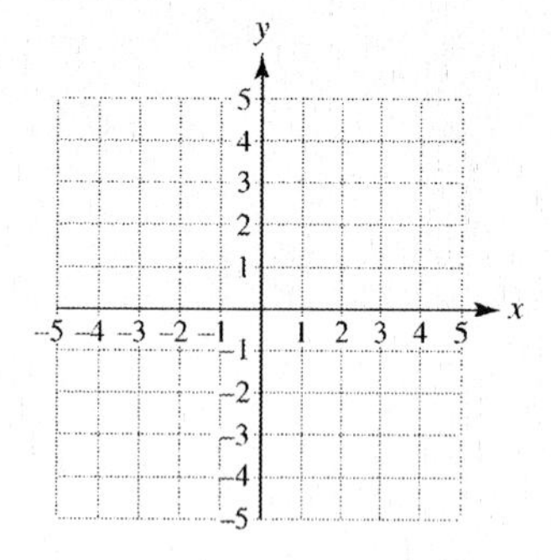

4. $y > -x + 5$

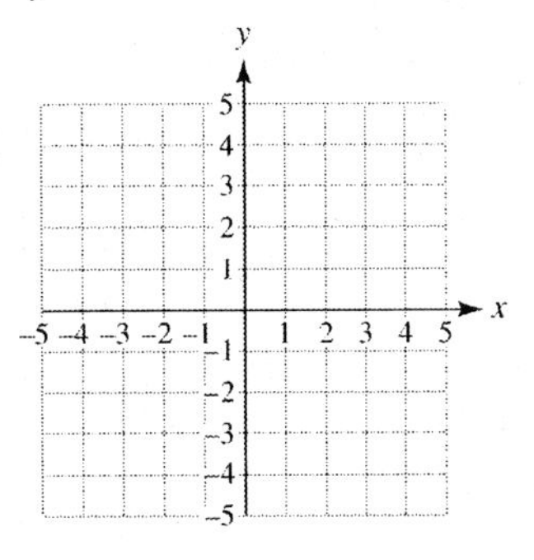

5. $2y - x \leq 4$

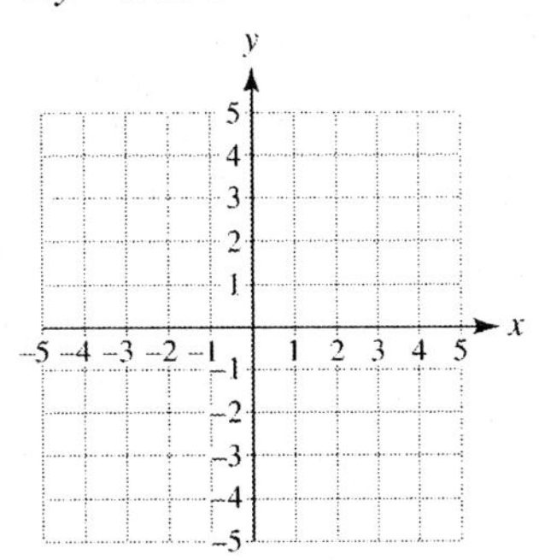

Solutions to Systems of Inequalities

Exercises 6-7: Refer to Examples 3-4 on pages 280-281 in your text and the Section 4.4 lecture video.

Shade the solution set to the system of inequalities.

6. $y \leq x$

$x + y > 2$

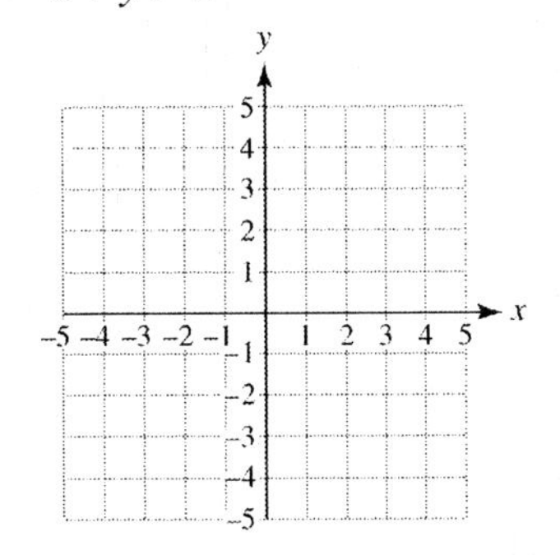

7. $x + 2y < -5$

$2x - y \leq 4$

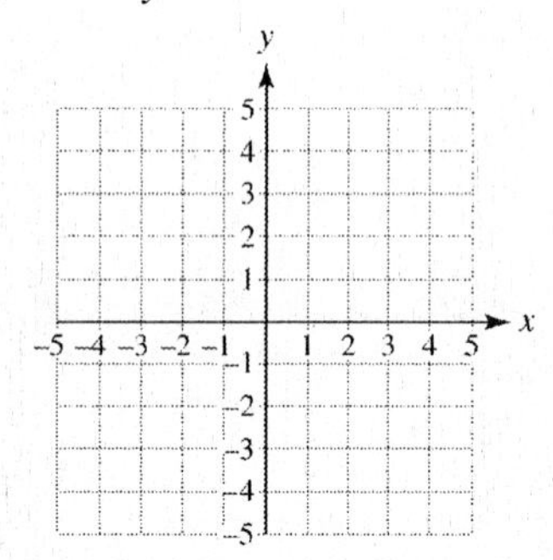

Applications

Exercises 8-9: Refer to Examples 5-6 on pages 282-283 in your text and the Section 4.4 lecture video.

8. A business manufactures at least two MP3 players for every CD player. The total number of MP3 players and CD players is no more than 120. Shade the region that shows the number of MP3 players M and CD players P that can be produced within these restrictions. Label the horizontal axis P and the vertical axis M.

8. 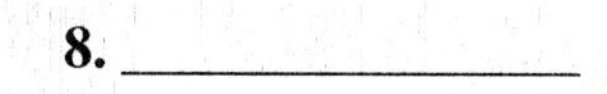

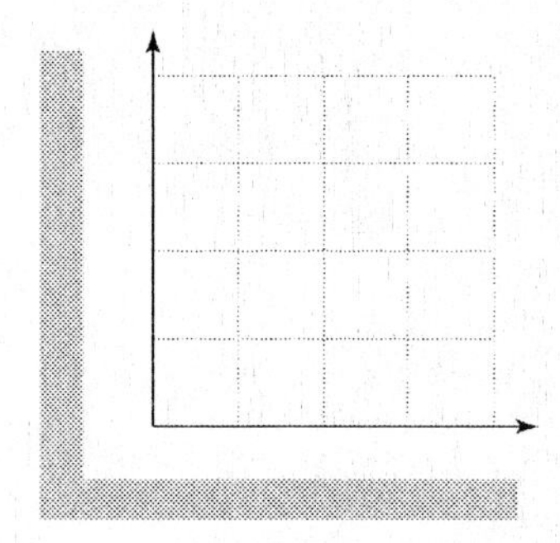

9. The figure shows a shaded region containing recommended weights w for heights h.

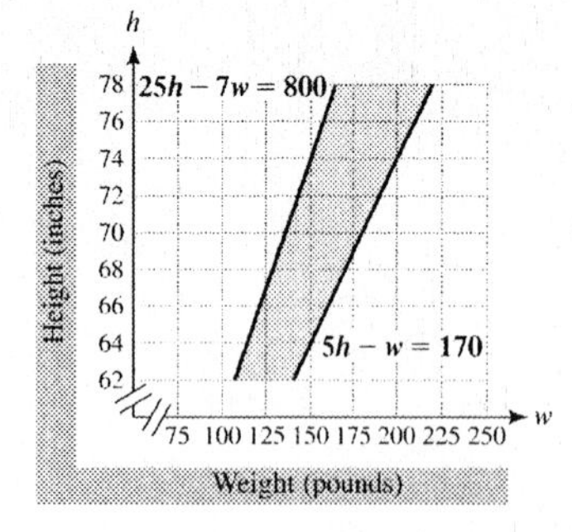

(a) What does the graph indicate about a 72-inch person who weighs 200 pounds?

9.(a)______________

(b) Determine the range of recommended weights for someone who is 74 inches tall.

(b)______________

Name: **Course/Section:** **Instructor:**

Chapter 5 Polynomials and Exponents
5.1 Rules for Exponents

Review of Bases and Exponents ~ Zero Exponents ~ The Product Rule ~ Power Rules

STUDY PLAN

Read: Read Section 5.1 on pages 298-304 in your textbook or eText.

Practice: Do your assigned exercises in your ☐ Book ☐ MyMathLab ☐ Worksheets

Review: Keep your corrected assignments in an organized notebook and use them to review for the test.

Key Terms

Exercises 1-6: Use the vocabulary terms listed below to complete each statement. Note that some terms or expressions may not be used.

exponent	**undefined**
power to a power	**product**
zero exponent	**product to a power**
base	**quotient to a power**

1. The ________________ rule for exponents states that for any real number a and natural numbers m and n, $\left(a^m\right)^n = a^{m \cdot n}$.

2. The ________________ rule for exponents states that for any real numbers a and b and natural number n, $\left(\frac{a}{b}\right)^n = \frac{a^n}{b^n}$, $b \neq 0$.

3. The ________________ rule for exponents states that for any real number a and natural numbers m and n, $a^m \cdot a^n = a^{m+n}$.

4. The exponential expression 3^4 has ________________ 3 and ________________ 4.

5. The ________________ rule for exponents states that for any real numbers a and b and natural number n, $(ab)^n = a^n b^n$.

6. The ________________ rule states that for any nonzero real number b, $b^0 = 1$.
   The expression 0^0 is ________________.

Review of Bases and Exponents

Exercises 1-4: Refer to Example 1 on pages 298-299 in your text and the Section 5.1 lecture video.

Evaluate each expression.

1. $3+\frac{2^3}{2}$ **1.** ______________

2. $2\left(\frac{1}{2}\right)^3$ **2.** ______________

3. -3^4 **3.** ______________

4. $(-4)^2$ **4.** ______________

Zero Exponents

Exercises 5-7: Refer to Example 2 on page 300 in your text and the Section 5.1 lecture video.

Evaluate each expression. Assume that all variables represent nonzero numbers.

5. -3^0 **5.** ______________

6. $4\left(\frac{2}{3}\right)^0$ **6.** ______________

7. $\left(\frac{xy^3}{2z}\right)^0$ **7.** ______________

The Product Rule

Exercises 8-13: Refer to Examples 3-4 on pages 300-301 in your text and the Section 5.1 lecture video.

Multiply and simplify.

8. $3^2 \cdot 3^3$ **8.** ____________

9. x^6x^2 **9.** ____________

10. $3x^4 \cdot 4x^3$ **10.** ____________

11. $x^4\left(3x^2+2x\right)$ **11.** ____________

12. $a \cdot a^4$ **12.** ____________

13. $(x+y)^3(x+y)$ **13.** ____________

Power Rules

Exercises 14-26: Refer to Examples 5-9 on pages 301-304 in your text and the Section 5.1 lecture video.

Simplify each expression.

14. $\left(2^3\right)^5$ **14.** ______________

15. $\left(x^4\right)^4$ **15.** ______________

16. $(5t)^3$ **16.** ______________

17. $\left(-2b^3\right)^3$ **17.** ______________

18. $3\left(x^3y^2\right)^4$ **18.** ______________

19. $\left(-2^2z^6\right)^3$ **19.** ______________

20. $\left(\frac{3}{4}\right)^3$ **20.** ______________

21. $\left(\frac{x}{y}\right)^6$

21. ________________

22. $\left(\frac{6}{a-b}\right)^2$

22. ________________

23. $(5x)^2(2x)^3$

23. ________________

24. $\left(\frac{a^3b}{c^2}\right)^3$

24. ________________

25. $(3xy^3)^2(-2x^4y^2)^3$

25. ________________

26. If a parcel of property increases in value by about 16% each year for 15 years, then its value will triple two times.

(a) Write an exponential expression that represents "tripling two times."

(b) If the property is initially worth $105,000, how much will it be worth if it triples 2 times?

26. (a) ________________

(b) ________________

Name: **Course/Section:** **Instructor:**

Chapter 5 Polynomials and Exponents
5.2 Addition and Subtraction of Polynomials

Monomials and Polynomials ~ Addition of Polynomials ~ Subtraction of Polynomials ~ Evaluating Polynomial Expressions

STUDY PLAN

Read: Read Section 5.2 on pages 306-313 in your textbook or eText.

Practice: Do your assigned exercises in your ☐ Book ☐ MyMathLab ☐ Worksheets

Review: Keep your corrected assignments in an organized notebook and use them to review for the test.

Key Terms

Exercises 1-10: Use the vocabulary terms listed below to complete each statement. Note that some terms or expressions may not be used. Some terms may be used more than once.

like	**unlike**
degree	**monomial**
binomial	**coefficient**
polynomial	**trinomial**

1. A polynomial with three terms is called a(n) ________________.

2. The ________________ of a monomial is the sum of the exponents of the variables.

3. A(n) ________________ is a monomial or the sum of two or more monomials.

4. If two monomials contain the same variables raised to the same powers, we call them ________________ terms.

5. A(n) ________________ is a number, a variable, or a product of numbers and variables raised to natural number powers.

6. The expression $4x^3 - 3x^2 + 5x - 8$ is a(n) ________________ in one variable.

7. The terms $7a^2b$ and $10ab^2$ are ________________ terms.

8. The number in a monomial is called the ________________ of the monomial.

9. A polynomial with two terms is called a(n) ________________.

10. The ________________ of a polynomial is the degree of the term (or monomial) with highest degree.

Monomials and Polynomials

***Exercises 1-3:** Refer to Example 1 on pages 307-308 in your text and the Section 5.2 lecture video.*

Determine whether the expression is a polynomial. If it is, state how many terms and variables the polynomial contains and give its degree.

1. $2x^3 + 4x - 3$ **1.** ______________

2. $4x^3 - 2x^2y^3 - xy^2 + 4y^3$ **2.** ______________

3. $3x^2 + \dfrac{1}{x-3}$ **3.** ______________

Addition of Polynomials

***Exercises 4-10:** Refer to Examples 2-5 on pages 309-310 in your text and the Section 5.2 lecture video.*

State whether each pair of expressions contains like terms or unlike terms. If they are like terms, add them.

4. $5b^3;\ 2b^2$ **4.** ______________

5. $-3x^3y;\ 7x^3y$ **5.** ______________

6. $3xy^2;\ \dfrac{1}{2}xy^2$ **6.** ______________

Add by combining like terms.

7. $(2x+1)+(-3x+5)$ **7.** ______________

8. $(4y^2 - 3y + 7) + (y^2 + 6y + 1)$ **8.** ______________

9. Add $(7x^3 - 6x^2 + 2x - 4) + (5x^3 - 4x^2 + 3)$ by combining like terms.

9. ______________

10. Simplify $(6b^2 + b + 7) + (-2b^2 - 5b + 6)$.

10. ______________

Subtraction of Polynomials

Exercises 11-14: Refer to Examples 6-7 on page 311 in your text and the Section 5.2 lecture video.

Simplify each expression.

11. $(x + 5) - (-2x + 7)$

11. ______________

12. $(4x^2 - 3x + 2) - (5x^2 + 7x + 1)$

12. ______________

13. $(x^3 + 5x^2 - 8) - (-3x^3 + 6x - 4)$

13. ______________

14. Simplify $(8x^2 - 3x - 1) - (5x^2 - 4x + 5)$.

14. ______________

Evaluating Polynomial Expressions

Exercises 15-16: Refer to Examples 8-9 on pages 312-313 in your text and the Section 5.2 lecture video.

15. Write the monomial that represents the volume of a box having length x inches, width y inches, and height 4 inches. Find the volume of the box if $x = 8$ in. and $y = 10$ in.

15. ______________

____

16. The polynomial $0.835x - 1635$ approximates the price of a first-class postage stamp (in cents), where $x = 1980$ corresponds to the year 1980, $x = 1981$ to the year 1981, and so on. Estimate the price of a first-class postage stamp in the year 2002. Round answer to the nearest cent.

16. ______________

Name: Course/Section: Instructor:

Chapter 5 Polynomials and Exponents
5.3 Multiplication of Polynomials

Multiplying Monomials ~ Review of the Distributive Properties ~ Multiplying Monomials and Polynomials ~ Multiplying Polynomials

STUDY PLAN

Read: Read Section 5.3 on pages 316-322 in your textbook or eText.

Practice: Do your assigned exercises in your ☐ Book ☐ MyMathLab ☐ Worksheets

Review: Keep your corrected assignments in an organized notebook and use them to review for the test.

Key Terms

Exercise 1-3: Use the vocabulary terms listed below to complete each statement. Note that some terms or expressions may not be used. Some terms may be used more than once.

binomial	**polynomial**
variable	**like term**
product	**coefficient**

1. To multiply monomials in one variable, do the following.
Step 1: Multiply the ________________(s).
Step 2: Use the product rule for exponents to multiply the ________________(s).
Step 3: Write the above results as a(n) ________________.

2. The ________________ of two polynomials may be found by multiplying every term in the first polynomial by every term in the second polynomial and then combining ________________(s).

3. The FOIL process may be helpful for remembering how to multiply two ________________(s).

Multiplying Monomials

Exercises 1-2: Refer to Example 1 on page 317 in your text and the Section 5.3 lecture video.

Multiply.

1. $-4x^3 \cdot 3x^2$ **1.** ______________

2. $(4a^3b^2)(-5ab^4)$ **2.** ______________

Review of the Distributive Properties

Exercises 3-5: Refer to Example 2 on page 317 in your text and the Section 5.3 lecture video.

Multiply.

3. $3(2x-1)$ **3.** ______________

4. $(4x^2-5)2$ **4.** ______________

5. $-x(5x-2)$ **5.** ______________

Multiplying Monomials and Polynomials

Exercises 6-11: Refer to Examples 3-4 on page 318 in your text and the Section 5.3 lecture video.

Multiply.

6. $5x(3x^2-4)$ **6.** ______________

7. $(4x-1)x^2$ **7.** ______________

8. $-2\left(3x^2-4x+2\right)$ **8.** ______________

9. $3x^3\left(x^3-2x^2+3\right)$ **9.** ______________

10. $4xy\left(5x^3y^2+1\right)$ **10.** ______________

11. $-xy\left(x^2+y^2\right)$ **11.** ______________

Multiplying Polynomials

Exercises 12-20: Refer to Examples 5-9 on pages 319-321 in your text and the Section 5.3 lecture video.

12. Multiply $(x+3)(x+5)$ geometrically and symbolically. **12.** ______________

Multiply. Draw arrows to show how each term is found.

13. $(2x-3)(x-1)$ **13.** ______________

14. $(3-x)(1+4x)$ **14.** ______________

15. $(3x-2)\left(x^2+x\right)$ **15.** ______________

Multiply.

16. $(4x+3)(x^2-x+1)$ **16.** ____________

17. $(a+b)(4a^2-5b^2)$ **17.** ____________

18. $(b^4-3b^2+1)(b^2-1)$ **18.** ____________

19. Multiply x^2+3x+4 and $x-2$ vertically. **19.** ____________

20. A box has a width 2 inches less than its height and a length 5 inches more than its height.

(a) If h represents the height of the box, write a polynomial that represents the volume of the box. **20.(a)**____________

(b) Use this polynomial to calculate the volume of the box if $h=15$ inches. **(b)**____________

Name: **Course/Section:** **Instructor:**

Chapter 5 Polynomials and Exponents
5.4 Special Products

Product of a Sum and Difference ~ Squaring Binomials ~ Cubing Binomials

STUDY PLAN

Read: Read Section 5.4 on pages 324-329 in your textbook or eText.

Practice: Do your assigned exercises in your ☐ Book ☐ MyMathLab ☐ Worksheets

Review: Keep your corrected assignments in an organized notebook and use them to review for the test.

Key Terms

Exercise 1-3: Use the expressions listed below to complete each statement. Note that some expressions may not be used.

a^2+b^2
a^2-b^2
a^2+ab+b^2
$a^2+2ab+b^2$
a^2-ab+b^2
$a^2-2ab+b^2$

1. For any real numbers a and b, $(a+b)(a-b)=$ ________________.

2. For any real numbers a and b, $(a+b)^2=$ ________________.

3. For any real numbers a and b, $(a-b)^2=$ ________________.

Product of a Sum and Difference

Exercises 1-5: Refer to Examples 1-2 on pages 325-326 in your text and the Section 5.4 lecture video.

Multiply.

1. $(a-1)(a+1)$ **1.** ______________

2. $(z+3)(z-3)$ **2.** ______________

3. $(3r+5t)(3r-5t)$ **3.** ______________

4. $(3m^2+2n^2)(3m^2-2n^2)$ **4.** ______________

5. Use the product of a sum and difference to find $23 \cdot 17$. **5.** ______________

Squaring Binomials

Exercises 6-10: Refer to Examples 3-4 on pages 326-327 in your text and the Section 5.4 lecture video.

Multiply.

6. $(x-4)^2$ **6.** ______________

7. $(3x-2)^2$ **7.** ______________

8. $(1+6b)^2$ **8.** ____________

9. $(3a^2-b)^2$ **9.** ____________

10. A square pool has a 6-foot-wide walk around it.

(a) If the sides of the pool have length x, find a polynomial that gives the total area of the pool and sidewalk. **10. (a)**____________

(b) Let $x = 20$ ft and evaluate the polynomial. **(c)**____________

Cubing Binomials

Exercises 11-12: Refer to Examples 5-6 on pages 328-329 in your text and the Section 5.4 lecture video.

11. Multiply $(5x+2)^3$. **11.** ____________

12. If a savings account pays r percent annual interest, where r is expressed as a decimal, then after 3 years a sum of money will grow by a factor of $(1+r)^3$.

(a) Multiply this expression. **12. (a)**____________

(b) Evaluate the expression for $r = 0.04$ (or 4%), and interpret the result. Round answer to the nearest thousandth. **(b)**____________

Name: **Course/Section:** **Instructor:**

Chapter 5 Polynomials and Exponents
5.5 Integer Exponents and the Quotient Rule

Negative Integers as Exponents ~ The Quotient Rule ~ Other Rules for Exponents ~ Scientific Notation

STUDY PLAN

Read: Read Section 5.5 on pages 332-340 in your textbook or eText.

Practice: Do your assigned exercises in your ☐ Book ☐ MyMathLab ☐ Worksheets

Review: Keep your corrected assignments in an organized notebook and use them to review for the test.

Key Terms

Exercise 1-4: Use the vocabulary terms listed below to complete each statement. Note that some terms or expressions may not be used.

$a^m - a^n$	$\left(\frac{b}{a}\right)^n$	a^{m-n}	$\frac{b^m}{a^n}$
$\frac{1}{a^n}$	$\frac{a^n}{b^m}$	$\left(\frac{a}{b}\right)^n$	a^n
reciprocal		**scientific notation**	

1. Let a be a nonzero real number and n be a positive integer. Then a^{-n} = ________________.
   That is, a^{-n} is the ________________ of a^n.

2. For any nonzero number a and integers m and n, $\frac{a^m}{a^n}$ = ________________.

3. The following three rules hold for any nonzero numbers a and b and positive integers m and n.
   (a) $\frac{1}{a^{-n}}$ = ________________.
   (b) $\frac{a^{-n}}{b^{-m}}$ = ________________.
   (c) $\left(\frac{a}{b}\right)^{-n}$ = ________________.

4. A real number a is in ________________ when a is written in the form $b \times 10^n$, where $1 \le |b| < 10$ and n is an integer.

Negative Integers as Exponents

Exercises 1-10: Refer to Examples 1-3 on pages 333-334 in your text and the Section 5.5 lecture video.

Simplify each expression.

1. 4^{-2} **1.** ______________

2. 5^{-1} **2.** ______________

3. x^{-3} **3.** ______________

4. $(a+b)^{-4}$ **4.** ______________

Evaluate each expression.

5. $3^2 \cdot 3^{-5}$ **5.** ______________

6. $5^{-3} \cdot 5^{-1}$ **6.** ______________

Simplify the expression. Write the answer using positive exponents.

7. $x^3 \cdot x^{-5}$ **7.** ______________

8. $(t^3)^{-2}$ **8.** ______________

9. $(ab)^{-4}$ **9.** ______________

10. $(xy)^{-5}(xy^{-3})^2$ **10.** ______________

The Quotient Rule

Exercises 11-13: Refer to Example 4 on page 335 in your text and the Section 5.5 lecture video.

Simplify each expression. Write the answer using positive exponents.

11. $\frac{3^4}{3^6}$ **11.** ________________

12. $\frac{15a^2}{5a^6}$ **12.** ________________

13. $\frac{x^6 y^2}{x^5 y^7}$ **13.** ________________

Other Rules for Exponents

Exercises 14-17: Refer to Example 5 on page 336 in your text and the Section 5.5 lecture video.

Simplify each expression. Write the answer using positive exponents.

14. $\frac{1}{3^{-4}}$ **14.** ________________

15. $\frac{4^{-2}}{3^{-3}}$ **15.** ________________

16. $\frac{4x^2 y^{-3}}{12x^{-3} y^4}$ **16.** ________________

17. $\left(\frac{b^3}{5}\right)^{-2}$ **17.** ________________

Scientific Notation

Exercises 18-23: Refer to Examples 6-8 on pages 338-339 in your text and the Section 5.5 lecture video.

Write each number in standard form.

18. 4.17×10^5 **18.** ______________

19. 2.3×10^{-4} **19.** ______________

20. 5.82×10^{-2} **20.** ______________

Write each number in scientific notation.

21. 32,000,000 **21.** ______________

22. 0.00002 **22.** ______________

23. There are approximately 5.859×10^{12} miles in one light year. The distance from Earth to the moon and back is 4.8×10^5 miles. How many round trips to the moon does one light-year represent? Write your answer in scientific notation. **23.** ______________

Name: Course/Section: Instructor:

Chapter 5 Polynomials and Exponents
5.6 Division of Polynomials

Division by a Monomial ~ Division by a Polynomial

STUDY PLAN

Read: Read Section 5.6 on pages 343-348 in your textbook or eText.

Practice: Do your assigned exercises in your ☐ Book ☐ MyMathLab ☐ Worksheets

Review: Keep your corrected assignments in an organized notebook and use them to review for the test.

Division by a Monomial

Exercises 1-5: Refer to Examples 1-3 on pages 343-345 in your text and the Section 5.6 lecture video.

Divide.

1. $\dfrac{x^6 - x^3}{x^2}$

1. ____________

2. $\dfrac{4b^8 - 12b^5}{8b^3}$

2. ____________

3. $\dfrac{15x^2 - 5x + 10}{5x}$

3. ____________

4. Divide the expression $\dfrac{6a^3+8a^2-4a}{2a^2}$ and then check the result. **4.** ______________

5. A rectangle has an area $A=x^2+3x$ and width x. Write an expression for its length l in terms of x. **5.** ______________

Division by a Polynomial

Exercises 6-8: Refer to Examples 4-6 on pages 345-347 in your text and the Section 5.6 lecture video.

6. Divide $\dfrac{6x^2+5x-8}{2x+3}$ and check. **6.** ______________

7. Simplify $\left(4x^3-2x+5\right)\div\left(x-1\right)$. **7.** ______________

8. Divide $x^3-4x^2+5x-12$ by x^2+4. **8.** ______________

Name: **Course/Section:** **Instructor:**

Chapter 6 Factoring Polynomials and Solving Equations
6.1 Introduction to Factoring

Basic Concepts ~ Common Factors ~ Factoring by Grouping

STUDY PLAN

Read: Read Section 6.1 on pages 360-366 in your textbook or eText.

Practice: Do your assigned exercises in your ☐ Book ☐ MyMathLab ☐ Worksheets

Review: Keep your corrected assignments in an organized notebook and use them to review for the test.

Key Terms

Exercises 1-3: Use the vocabulary terms listed below to complete each statement. Note that some terms or expressions may not be used.

factor
completely factored
greatest common factor (GCF)

1. We _______________ a polynomial by writing the polynomial as a product of two or more lower degree polynomials.

2. The _______________ of a list of positive integers is the largest integer that is a factor of every integer in the list.

3. A term is _______________ when its coefficient is written as a product of prime numbers and any powers of variables are written as repeated multiplication.

Common Factors

Exercises 1-10: Refer to Examples 1-4 on pages 361-363 in your text and the Section 6.1 lecture video.

Factor the expression and sketch a rectangle that illustrates the factorization.

1. $30x+18$ **1.** ____________

2. $9x^2-12x$ **2.** ____________

Factor.

3. $12x^2+8x$ **3.** ____________

4. $10y^3-2y^2$ **4.** ____________

5. $4a^3-12a^2-8a$ **5.** ____________

6. $6x^3y-2x^2y^2$ **6.** ____________

Find the greatest common factor for each expression. Then factor the expression.

7. $4a^2 + 6a$

7. ________________

____

8. $5x^4 - 15x^2$

8. ________________

____

9. $6a^3b^2 - 15a^2b^3$

9. ________________

____

10. If a ball is thrown upward at 40 feet per second, then its height after t seconds is approximated by $40t - 16t^2$. Factor this expression.

10. ________________

Factoring by Grouping

Exercises 11-18: Refer to Examples 5-8 on pages 363-365 in your text and the Section 6.1 lecture video.

Factor.

11. $4x(x-3)-5(x-3)$

11. ________________

12. $t^2(2t+3)+6(2t+3)$

12. ________________

Factor each polynomial.

13. $3x^3 - 6x^2 + 4x - 8$ **13.** ______________

14. $4x - 4y + ax - ay$ **14.** ______________

15. $5x^3 - 15x^2 - x + 3$ **15.** ______________

16. $2z^3 + 12z^2 - 3z - 18$ **16.** ______________

Completely factor each polynomial.

17. $8x^3 - 12x^2 + 8x - 12$ **17.** ______________

18. $4x^5 + 6x^4 - 10x^3 - 15x^2$ **18.** ______________

Name: **Course/Section:** **Instructor:**

Chapter 6 Factoring Polynomials and Solving Equations

6.2 Factoring Trinomials I $(x^2 + bx + c)$

Review of the FOIL Method ~ Factoring Trinomials with Leading Coefficient 1

STUDY PLAN

Read: Read Section 6.2 on pages 368-373 in your textbook or eText.

Practice: Do your assigned exercises in your ☐ Book ☐ MyMathLab ☐ Worksheets

Review: Keep your corrected assignments in an organized notebook and use them to review for the test.

Key Terms

Exercises 1-2: Use the vocabulary terms listed below to complete each statement. Note that some terms or expressions may not be used.

standard form
prime polynomial
leading coefficient

1. Any trinomial of degree 2 in the variable x can be written in ________________ as $ax^2 + bx + c$, where a, b, and c are constants. The constant a is called the ________________.

2. A polynomial with integer coefficients that cannot be factored by using integer coefficients is called a(n) ________________.

Factoring Trinomials with Leading Coefficient 1

Exercises 1-16: Refer to Examples 1-7 on pages 369-373 in your text and the Section 6.2 lecture video.

For each of the following, find an integer pair that has the given product and sum.

1. Product: 28; Sum: 11 **1.** ______________

2. Product: -40; Sum: -3 **2.** ______________

Factor each trinomial.

3. $x^2+7x+10$ **3.** ______________

4. $x^2+9x+18$ **4.** ______________

5. $y^2+13y+42$ **5.** ______________

6. $b^2-10b+21$ **6.** ______________

7. $x^2-8x+12$ **7.** ______________

8. y^2-y-20 **8.** ______________

9. $t^2 - 3t - 40$ **9.** ____________

10. $x^2 + 2x - 24$ **10.** ____________

11. $x^2 - 7x + 12$ **11.** ____________

Factor each trinomial, if possible.

12. $x^2 - 9x + 22$ **12.** ____________

13. $x^2 - 5x - 14$ **13.** ____________

Factor each trinomial completely.

14. $5x^2 + 30x + 40$ **14.** ____________

15. $2x^4 + 10x^3 - 12x^2$ **15.** ____________

16. Find one possibility for the dimensions of a rectangle that has an area of $x^2 + 3x - 10$. **16.** ____________

Name: **Course/Section:** **Instructor:**

Chapter 6 Factoring Polynomials and Solving Equations
6.3 Factoring Trinomials II $(ax^2 + bx + c)$

Factoring Trinomials by Grouping ~ Factoring with FOIL in Reverse

STUDY PLAN

Read: Read Section 6.3 on pages 376-382 in your textbook or eText.

Practice: Do your assigned exercises in your ☐ Book ☐ MyMathLab ☐ Worksheets

Review: Keep your corrected assignments in an organized notebook and use them to review for the test.

Factoring Trinomials by Grouping

Exercises 1-7: Refer to Examples 1-3 on pages 377-379 in your text and the Section 6.3 lecture video.

Factor each trinomial.

1. $2x^2 + 13x + 15$ **1.** ______________

2. $4a^2 - 11a - 3$ **2.** ______________

3. $10x^2 - 29x + 10$ **3.** ______________

4. $5x^2 + 9x - 5$ **4.** ______________

5. $2a^2 - 9a - 10$ **5.** ______________

Factor each trinomial completely.

6. $15y^2 - 55y - 20$ **6.** ______________

7. $3x^3 - 18x^2 - 27x$ **7.** ______________

Factoring with FOIL in Reverse

Exercises 8-12: Refer to Examples 4-5 on pages 380-381 in your text and the Section 6.3 lecture video.

Factor each trinomial.

8. $3a^2 - 11a - 20$ **8.** ______________

9. $2x^2 + 11x + 5$ **9.** ______________

10. $8 + 5x^2 + 41x$ **10.** ______________

11. $-6x^2 - 13x + 5$ **11.** ______________

12. $3x + 35 - 2x^2$ **12.** ______________

Name: ______________ **Course/Section:** ______________ **Instructor:** ______________

Chapter 6 Factoring Polynomials and Solving Equations
6.4 Special Types of Factoring

Difference of Two Squares ~ Perfect Square Trinomials ~ Sum and Difference of Two Cubes

STUDY PLAN

Read: Read Section 6.4 on pages 384-389 in your textbook or eText.

Practice: Do your assigned exercises in your ☐ Book ☐ MyMathLab ☐ Worksheets

Review: Keep your corrected assignments in an organized notebook and use them to review for the test.

Key Terms

Exercises 1-6: Use the expressions listed below to complete each statement. Note that some expressions may not be used.

$a^2 + b^2$ | $a^2 + 2ab + b^2$
$(a+b)^2$ | $a^2 - 2ab + b^2$
$a^2 - b^2$ | $(a-b)(a^2 + ab + b^2)$
$(a-b)^2$ | $(a-b)(a^2 + 2ab + b^2)$
$(a+b)^3$ | $(a+b)(a^2 - ab + b^2)$
$(a-b)^3$ | $(a+b)(a^2 - 2ab + b^2)$
$(a-b)(a+b)$

1. For any real numbers a and b, $a^3 + b^3 =$ ______________.

2. For any real numbers a and b, $a^2 - 2ab + b^2 =$ ______________.

3. For any real numbers a and b, $a^3 - b^3 =$ ______________.

4. For any real numbers a and b, $a^2 + 2ab + b^2 =$ ______________.

5. For any real numbers a and b, $a^2 - b^2 =$ ______________.

6. The expressions ______________ and ______________ are called perfect square trinomials.

Difference of Two Squares

***Exercises 1-4:** Refer to Example 1 on page 384 in your text and the Section 6.4 lecture video.*

Factor each difference of two squares.

1. $x^2 - 16$ **1.** ______________

2. $25x^2 - 4$ **2.** ______________

3. $81 - 25a^2$ **3.** ______________

4. $9x^2 - 100y^2$ **4.** ______________

Perfect Square Trinomials

***Exercises 5-8:** Refer to Example 2 on pages 385-386 in your text and the Section 6.4 lecture video.*

If possible, factor each trinomial as a perfect square trinomial.

5. $x^2 + 12x + 36$ **5.** ______________

6. $9t^2 + 6t + 1$ **6.** ______________

7. $25x^2 - 40x + 16$ **7.** ______________

8. $x^2 - 14xy + 49y^2$ **8.** ______________

Sum and Difference of Two Cubes

Exercises 9-16: Refer to Examples 3-5 on pages 387-388 in your text and the Section 6.4 lecture video.

Factor each polynomial.

9. x^3+125 **9.** ______________

10. a^3-8 **10.** ______________

11. x^3-216 **11.** ______________

12. $64x^3-27$ **12.** ______________

13. $16y^2+24y+9$ **13.** ______________

14. $16b^2-25$ **14.** ______________

Factor each polynomial completely.

15. $12x^3-60x^2+75x$ **15.** ______________

16. $9a^3-64ab^2$ **16.** ______________

Name: **Course/Section:** **Instructor:**

Chapter 6 Factoring Polynomials and Solving Equations
6.5 Summary of Factoring

Guidelines for Factoring Polynomials ~ Factoring Polynomials

STUDY PLAN

Read: Read Section 6.5 on pages 391-395 in your textbook or eText.

Practice: Do your assigned exercises in your ☐ Book ☐ MyMathLab ☐ Worksheets

Review: Keep your corrected assignments in an organized notebook and use them to review for the test.

Key Terms

Exercises: Use the vocabulary terms listed below to complete each statement. Note that some terms or expressions may not be used. Some terms may be used more than once.

$a^2 + b^2$
$(a+b)^2$
$a^2 - b^2$
$(a-b)^2$
$a^3 + b^3$
$a^3 - b^3$

FOIL
grouping
perfect square
sum of two cubes
completely factored
difference of two squares
perfect square trinomial
greatest common factor (GCF)
difference of two cubes

Guidelines for Factoring Polynomials

STEP 1: Factor out the ________________, if possible.

STEP 2: **A.** If the polynomial has four terms, try factoring by ________________.

B. If the polynomial is a binomial, try one of the following.

1. ________________ $= (a-b)(a+b)$ This is referred to as a(n) ________________.

2. ________________ $= (a-b)(a^2+ab+b^2)$ This is referred to as a(n) ________________.

3. ________________ $= (a+b)(a^2-ab+b^2)$ This is referred to as a(n) ________________.

C. If the polynomial is a trinomial, check for a(n) ________________.

1. $a^2 + 2ab + b^2 =$ ________________ This is referred to as a(n) ________________.

2. $a^2 - 2ab + b^2 =$ ________________ This is referred to as a(n) ________________.

Otherwise, try to factor the trinomial by ________________ or apply ________________ in reverse.

STEP 3: Check to make sure that the polynomial is ________________.

Factoring Polynomials

Exercises 1-8: Refer to Examples 1-8 on pages 392-394 in your text and the Section 6.5 lecture video.

Factor.

1. $5x^3 - 20x^2 + 25x$ **1.** ______________

2. $4t^4 - 144t^2$ **2.** ______________

3. $-45a^3 - 30a^2 - 5a$ **3.** ______________

4. $5x^3 - 320$ **4.** ______________

5. $24x^4 + 10x^3 - 4x^2$ **5.** ______________

6. $8x^3 + 4x^2 - 72x - 36$ **6.** ______________

7. $16a^3b - 36ab^3$ **7.** ______________

8. $12x^3 + 9x^2 + 20x + 15$ **8.** ______________

Name: **Course/Section:** **Instructor:**

Chapter 6 Factoring Polynomials and Solving Equations
6.6 Solving Equations by Factoring I (Quadratics)

The Zero-Product Property ~ Solving Quadratic Equations ~ Applications

STUDY PLAN

Read: Read Section 6.6 on pages 396-402 in your textbook or eText.

Practice: Do your assigned exercises in your ☐ Book ☐ MyMathLab ☐ Worksheets

Review: Keep your corrected assignments in an organized notebook and use them to review for the test.

Key Terms

Exercises 1-5: Use the vocabulary terms listed below to complete each statement. Note that some terms or expressions may not be used.

zeros
standard form
zero-product
quadratic equation
quadratic polynomial

1. The ________________ property states that if the product of two numbers (or expressions) is 0, then at least one of the numbers (or expressions) must equal 0.

2. Any ________________ in the variable x can be written as $ax^2 + bx + c$ with $a \neq 0$.

3. The ________________ of a polynomial in x are the values that, when substituted for x, result in 0.

4. Any ________________ in the variable x can be written as $ax^2 + bx + c = 0$ with $a \neq 0$.

5. The form $ax^2 + bx + c = 0$ is called the ________________ of a quadratic equation.

The Zero-Product Property

Exercises 1-4: Refer to Example 1 on page 397 in your text and the Section 6.6 lecture video.

Solve each equation.

1. $x(x+2)=0$ **1.** ______________

2. $3a^2=0$ **2.** ______________

3. $(b+1)(b-4)=0$ **3.** ______________

4. $x(x-3)(x+5)=0$ **4.** ______________

Solving Quadratic Equations

Exercises 5-9: Refer to Examples 2-3 on pages 398-399 in your text and the Section 6.6 lecture video.

Solve each quadratic equation. Check your answers.

5. $x^2+4x=0$ **5.** ______________

6. $t^2=9$ **6.** ______________

7. $a^2-5a+6=0$ **7.** ______________

8. $10x^2 + 7x = 12$ **8.** ______________

9. Solve $2x^2 - 9x = 5$. **9.** ______________

Applications

Exercises 10-12: Refer to Examples 4-6 on pages 400-401 in your text and the Section 6.6 lecture video.

10. The height h in feet of a baseball after t seconds is given by $h(t) = -16t^2 + 88t + 4$. At what values of t is the height of the baseball 100 feet? **10.** ______________

11. The braking distance D in feet required to stop a car traveling at x miles per hour on wet, level pavement can be approximated by $D = \frac{1}{9}x^2$.

(a) Calculate the braking distance for a car traveling at 40 miles per hour. **11.(a)**______________

(b) If the braking distance is 60 feet, estimate the speed of the car. **(b)**______________

(c) If you have a calculator available, use it to solve part (b) numerically with a table of values. **(c)**______________

12. A digital photograph is 20 pixels longer than it is wide and has a total area of 3500 pixels. Find the dimensions of this photograph. **12.** ______________

Understanding Concepts through Multiple Approaches
(For additional practice, visit MyMathLab.)

13. ***Braking Distance*** The braking distance D in feet required to stop a car traveling x miles per hour on wet, level pavement is approximated by $d = \frac{1}{9}x^2$. If the braking distance is 144 feet, estimate the speed of the car.

(a) Solve algebraically.

(b) Solve numerically using a calculator.

Did you get the same result using each method? Which method do you prefer? Explain why.

Name: **Course/Section:** **Instructor:**

Chapter 6 Factoring Polynomials and Solving Equations
6.7 Solving Equations by Factoring II (Higher Degree)

Polynomials with Common Factors ~ Special Types of Polynomials

STUDY PLAN

Read: Read Section 6.6 on pages 405-408 in your textbook or eText.

Practice: Do your assigned exercises in your ☐ Book ☐ MyMathLab ☐ Worksheets

Review: Keep your corrected assignments in an organized notebook and use them to review for the test.

Polynomials with Common Factors

Exercises 1-5: Refer to Examples 1-3 on pages 405-407 in your text and the Section 6.7 lecture video.

Factor each trinomial completely.

1. $-3x^2+9x+12$ **1.** ________________

2. $2x^3-12x^2+10x$ **2.** ________________

Solve each equation.

3. $6y^3-y^2-y=0$ **3.** ________________

4. $4x^3-4x^2=120x$ **4.** ________________

5. A box is made by cutting a square with sides of length x from each corner of a rectangular piece of metal with length 30 inches and width 20 inches. The box has no top. If the outside surface area of the box is 500 square inches, find the value of x. **5.** ______________

Special Types of Polynomials

Exercises 6-10: Refer to Examples 4-5 on pages 407-408 in your text and the Section 6.7 lecture video.

Factor each polynomial completely.

6. $x^4 - 81$ **6.** ______________

7. $a^4 + 6a^2 + 5$ **7.** ______________

8. $r^4 - 2r^2t^2 + t^4$ **8.** ______________

9. $a^4 - 16b^4$ **9.** ______________

10. $x^4 - 64x$ **10.** ______________

Name: **Course/Section:** **Instructor:**

Chapter 7 Rational Expressions
7.1 Introduction to Rational Expressions

Basic Concepts ~ Simplifying Rational Expressions ~ Applications

STUDY PLAN

Read: Read Section 7.1 on pages 420-427 in your textbook or eText.

Practice: Do your assigned exercises in your ☐ Book ☐ MyMathLab ☐ Worksheets

Review: Keep your corrected assignments in an organized notebook and use them to review for the test.

Key Terms

Exercises 1-6: Use the vocabulary terms listed below to complete each statement. Note that some terms or expressions may not be used.

undefined
lowest terms
basic principle
vertical asymptote
probability
defined
rational expression

1. An expression is written in ________________ when the numerator and denominator have no common factors.

2. The ________________ of an event indicates the likelihood that the event will occur.

3. Division by 0 is ________________.

4. A(n) ________________ indicates a value of x at which a rational expression is undefined.

5. A(n) ________________ can be written as $\frac{P}{Q}$, where P and Q are polynomials. It is ________________ whenever $Q \neq 0$.

6. The ________________ of rational expressions states that $\frac{P \cdot R}{Q \cdot R} = \frac{P}{Q}$ for $Q, R \neq 0$.

Basic Concepts

Exercises 1-8: Refer to Examples 1-2 on pages 420-421 in your text and the Section 7.1 lecture video.

If possible, evaluate each expression for the given value of the variable.

1. $\frac{1}{x-3}$ $\quad x=2$ **1.** ______________

2. $\frac{y^2}{3y-2}$ $\quad y=-2$ **2.** ______________

3. $\frac{3a+7}{a^2+2a+1}$ $\quad a=-1$ **3.** ______________

4. $\frac{x-4}{4-x}$ $\quad x=1$ **4.** ______________

Find all values of the variable for which each expression is undefined.

5. $\frac{1}{t}$ **5.** ______________

6. $\frac{5x}{3x+2}$ **6.** ______________

7. $\frac{1+4x}{x^2-9}$ **7.** ______________

8. $\frac{3}{a^2+4}$ **8.** ______________

Simplifying Rational Expressions

Exercises 9-17: Refer to Examples 3-5 on pages 421-424 in your text and the Section 7.1 lecture video.

Simplify each fraction by applying the basic principle of fractions.

9. $-\frac{8}{24}$ **9.** ________________

10. $\frac{25}{40}$ **10.** ________________

Simplify each expression.

11. $\frac{12t}{3t^2}$ **11.** ________________

12. $\frac{3x-6}{4x-8}$ **12.** ________________

13. $\frac{(x-2)(x+3)}{(x+3)(x-1)}$ **13.** ________________

14. $\frac{x^2-25}{2x^2-11x+5}$ **14.** ________________

15. $\frac{-t+4}{3t-12}$ **15.** ________________

16. $\dfrac{6-x}{x-6}$ **16.** ______________

17. $-\dfrac{7-a}{a-7}$ **17.** ______________

Applications

Exercises 18-20: Refer to Examples 6-8 on pages 424-426 in your text and the Section 7.1 lecture video.

18. Suppose that 8 cars per minute can pass through a construction zone. If traffic arrives randomly at an average rate of x cars per minute, the average time T in minutes spent waiting in line and passing through the construction zone is given by $T=\dfrac{1}{8-x}$, where $x<8$.

(a) Complete the table by finding T for each value of x.

x (cars/minute)	5	6	7	7.5	7.9	7.99
T (minutes)						

(b) Interpret the results. **18.(b)**______________

19. Suppose that a small fish species is introduced into a pond that had not previously held this type of fish, and that its population P in thousands is modeled by $P=\dfrac{2x+1}{x+5}$, where $x\geq 0$ represents time in months.

(a) Complete the table by finding P for each value of x. Round to 3 decimal places

x (months)	0	6	12	36	72
P (thousands)					

(b) How many fish were initially introduced into the pond? **19.(b)**______________

(c) Interpret the results shown in your completed table. **(c)**______________

20. Suppose that n balls, numbered 1 to n, are placed in a container and four balls have the winning number.

(a) What is the probability of drawing the winning ball at random?

20.(a)________________

(b) Calculate the probability for $n = 100, 1000,$ and $10,000$.

(b)________________

(c) What happens to the probability of drawing the winning ball as the number of balls increases?

(c)________________

Name: ______________________ **Course/Section:** ______________________ **Instructor:** ______________________

Chapter 7 Rational Expressions
7.2 Multiplication and Division of Rational Expressions

Review of Multiplication and Division of Fractions ~ Multiplication of Rational Expressions ~ Division of Rational Expressions

STUDY PLAN

Read: Read Section 7.2 on pages 431-434 in your textbook or eText.

Practice: Do your assigned exercises in your ☐ Book ☐ MyMathLab ☐ Worksheets

Review: Keep your corrected assignments in an organized notebook and use them to review for the test.

Key Terms

Exercises 1-4: Use the vocabulary terms listed below to complete each statement. Note that some terms or expressions may not be used.

reciprocal	**equivalent fraction**
numerator	**inverse principle of fractions**
lowest term	**basic principle of fractions**
denominator	**greatest common factor**

1. To multiply two rational expressions, multiply the ________________(s) and multiply the ________________(s).

2. The ________________ states that if the variables *a, b,* and *c* represent integers with $b \neq 0$ and $c \neq 0$, then $\frac{a \cdot c}{b \cdot c} = \frac{a}{b}$.

3. To divide two rational expressions, multiply by the ________________ of the divisor.

4. A fraction is simplified to ________________(s) if the GCF of its numerator and denominator is 1.

Review of Multiplication and Division of Fractions

Exercises 1-6: Refer to Examples 1-2 on page 431 in your text and the Section 7.2 lecture video.

Multiply and simplify your answers to lowest terms.

1. $\frac{2}{5}\cdot\frac{3}{7}$ **1.** ______________

2. $3\cdot\frac{5}{6}$ **2.** ______________

3. $\frac{5}{18}\cdot\frac{3}{20}$ **3.** ______________

Divide and simplify your answers to lowest terms.

4. $\frac{1}{5}\div\frac{3}{4}$ **4.** ______________

5. $\frac{3}{7}\div 9$ **5.** ______________

6. $\frac{7}{20}\div\frac{14}{15}$ **6.** ______________

Multiplication of Rational Expressions

Exercises 7-11: Refer to Examples 3-4 on pages 432-433 in your text and the Section 7.2 lecture video.

Multiply and simplify to lowest terms. Leave your answers in factored form.

7. $\frac{4}{x} \cdot \frac{2x+3}{x+1}$

7. ______________

8. $\frac{x+2}{x-5} \cdot \frac{3x}{x+2}$

8. ______________

9. $\frac{x^2-9}{3x-1} \cdot \frac{3x-1}{x-3}$

9. ______________

10. $\frac{5}{x^2-3x+2} \cdot \frac{x^2+3x-4}{10}$

10. ______________

11. If a car is traveling at 60 miles per hour on a slippery road, then its stopping distance D in feet can be calculated by

$$D = \frac{3600}{30} \cdot \frac{1}{x},$$

where x is the coefficient of friction between the tires and the road and $0 < x \leq 1$. The more slippery the road is, the smaller the value of x.

(a) Multiply and simplify the formula for D.

11.(a)______________

(b) Compare the stopping distance on an icy road with $x = 0.2$ to the stopping distance on dry pavement with $x = 0.6$.

(b)______________

Division of Rational Expressions

Exercises 12-14: Refer to Example 5 on pages 433-434 in your text and the Section 7.2 lecture video.

Divide and simplify to lowest terms.

12. $\frac{4}{3x} \div \frac{8}{x+2}$ **12.** ______________

13. $\frac{x^2-4}{x^2+1} \div (x+2)$ **13.** ______________

14. $\frac{x^2-3x}{x^2-2x-3} \div \frac{x}{x+3}$ **14.** ______________

Name: Course/Section: Instructor:

Chapter 7 Rational Expressions
7.3 Addition and Subtraction with Like Denominators

Review of Addition and Subtraction of Fractions ~ Rational Expressions with Like Denominators

STUDY PLAN

Read: Read Section 7.3 on pages 437-442 in your textbook or eText.

Practice: Do your assigned exercises in your ☐ Book ☐ MyMathLab ☐ Worksheets

Review: Keep your corrected assignments in an organized notebook and use them to review for the test.

Key Terms

Exercises 1-4: Use the vocabulary terms listed below to complete each statement. Note that some terms or expressions may not be used. Some terms may be used more than once.

greatest common factor
like denominator
equivalent fraction
denominator
numerator
add
subtract

1. If two rational expressions have the same denominator, we say that they have ________________(s).

2. To subtract two rational expressions having like denominators, ________________ their ________________(s). Keep the same ________________.

3. The largest number that divides evenly into two or more given numbers is known as the ________________.

4. To add two rational expressions having like denominators, ________________ their ________________(s). Keep the same ________________.

Review of Addition and Subtraction of Fractions

Exercises 1-4: Refer to Example 1 on page 437 in your text and the Section 7.3 lecture video.

Simplify each expression to lowest terms.

1. $\frac{2}{7}+\frac{3}{7}$ **1.** ______________

2. $\frac{3}{8}+\frac{3}{8}$ **2.** ______________

3. $\frac{9}{4}-\frac{5}{4}$ **3.** ______________

4. $\frac{17}{15}-\frac{7}{15}$ **4.** ______________

Rational Expressions with Like Denominators

Exercises 5-17: Refer to Examples 2-6 on pages 438-441 in your text and the Section 7.3 lecture video.

Add and simplify to lowest terms.

5. $\frac{4}{t}+\frac{5}{t}$ **5.** ______________

6. $\frac{x}{x+3}+\frac{3}{x+3}$ **6.** ______________

7. $\dfrac{a-2}{a^2+5a}+\dfrac{2}{a^2+5a}$

7. ________________

8. $\dfrac{x^2+x}{x+2}+\dfrac{3x+4}{x+2}$

8. ________________

9. $\dfrac{3}{ab}+\dfrac{4}{ab}$

9. ________________

10. $\dfrac{x}{x^2-y^2}+\dfrac{y}{x^2-y^2}$

10. ________________

11. $\dfrac{2}{a-b}+\dfrac{-2}{b-a}$

11. ________________

Subtract and simplify to lowest terms.

12. $\dfrac{x+3}{x}-\dfrac{3}{x}$

12. ________________

13. $\dfrac{3y}{2y-5}-\dfrac{5y}{2y-5}$

13. ________________

14. $\dfrac{x-4}{3x^2-x-4}-\dfrac{-5}{3x^2-x-4}$

14. ______________

15. $\dfrac{4x}{3x+4}-\dfrac{x}{3x+4}$

15. ______________

16. $\dfrac{a-b}{4b}-\dfrac{a+b}{4b}$

16. ______________

17. A container holds a mixtures of TI-84 and TI-89 calculators. In this container, there is a total of n calculators, including 5 defective TI-84 calculators and 8 defective TI-89 calculators. If a calculator is picked at random by a quality control inspector, then the probability, or chance, of one of the defective calculators being chosen is given by the expression $\dfrac{5}{n}+\dfrac{8}{n}$.

(a) Simplify this expression.

17.(a) ______________

(b) Interpret the result.

(b) ______________

Name: Course/Section: Instructor:

Chapter 7 Rational Expressions
7.4 Addition and Subtraction with Unlike Denominators

Finding Least Common Multiples ~ Review of Fractions with Unlike Denominators ~ Rational Expressions with Unlike Denominators

STUDY PLAN

Read: Read Section 7.4 on pages 444-451 in your textbook or eText.

Practice: Do your assigned exercises in your ☐ Book ☐ MyMathLab ☐ Worksheets

Review: Keep your corrected assignments in an organized notebook and use them to review for the test.

Key Terms

Exercises 1-5: Use the vocabulary terms listed below to complete each statement. Note that some terms or expressions may not be used.

listing method
greatest common factor
common multiple
least common multiple
prime factorization method
least common denominator

1. Used to find the LCM of two or more expressions, the ________________ involves finding the prime factorization of each expression. To find the LCM, list each factor the greatest number of times that it occurs in any one of the factorizations, and find the product of this list.

2. To add or subtract rational expressions with unlike denominators, we must first determine the ________________ for all rational expressions involved.

3. A(n) ________________ of two or more expressions is an expression that is divisible by each of the given expressions.

4. Used to find the LCM of two or more expressions, the ________________ involves writing the multiples of each expression and choosing the least common multiple of these.

5. The ________________ of two or more expressions is the simplest expression that is divisible by each of the expressions.

Finding Least Common Multiples

Exercises 1-5: Refer to Examples 1-2 on pages 444-445 in your text and the Section 7.4 lecture video.

Find the least common multiple of each pair of expressions.

1. $3a, 4a^2$ **1.** ________________

2. $z+1,\ z^2+z$ **2.** ________________

3. $x-4,\ x+5$ **3.** ________________

4. $x^2-3x+2,\ x^2+2x-3$ **4.** ________________

5. Use a step diagram to find the LCM of $3x^3-15x^2$ and $3x^3-12x^2-15x$. **5.** ________________

Review of Fractions with Unlike Denominators

Exercises 6-7: Refer to Example 3 on page 446 in your text and the Section 7.4 lecture video.

Simplify each expression.

6. $\frac{5}{12}-\frac{7}{18}$ **6.** ________________

7. $\frac{2}{7}+\frac{3}{5}$ **7.** ________________

Rational Expressions with Unlike Denominators

Exercises 8-17: Refer to Examples 4-7 on pages 446-450 in your text and the Section 7.4 lecture video.

Rewrite each rational expression so it has the given denominator D.

8. $\frac{5}{3x}$, $D = 9x^2$ **8.** ________________

9. $\frac{2}{x-y}$, $D = x^2 - y^2$ **9.** ________________

Find each sum and leave your answer in factored form.

10. $\frac{4}{9y} + \frac{5}{18y^2}$ **10.** ________________

11. $\frac{1}{x+2} + \frac{1}{x-2}$ **11.** ________________

12. $\frac{x}{x^2+6x+9} + \frac{1}{x+3}$ **12.** ________________

Simplify each expression. Write your answer in lowest terms and leave it in factored form.

13. $\frac{4}{a}-\frac{a}{a+2}$

13. ________________

14. $\frac{3}{x-1}-\frac{2}{x^2-1}$

14. ________________

15. $\frac{x}{x^2+x}-\frac{1}{x^2-x}$

15. ________________

16. $\frac{4}{x}-\frac{1}{x+3}-\frac{3}{x^2+3x}$

16. ________________

17. The length of a rectangle is $\frac{1}{x-5}$ and the width is $\frac{1}{x+1}$. Write an expression for the perimeter of the rectangle as a single rational expression in lowest terms.

17. ________________

Name: ______ **Course/Section:** ______ **Instructor:** ______

Chapter 7 Rational Expressions
7.5 Complex Fractions

Basic Concepts ~ Simplifying Complex Fractions

STUDY PLAN

Read: Read Section 7.5 on pages 454-460 in your textbook or eText.

Practice: Do your assigned exercises in your ☐ Book ☐ MyMathLab ☐ Worksheets

Review: Keep your corrected assignments in an organized notebook and use them to review for the test.

Key Terms

Exercises 1-4: Use the vocabulary terms listed below to complete each statement. Note that some terms or expressions may not be used.

reciprocal
complex fraction
improper fraction
basic complex fraction
greatest common factor
least common denominator

1. A(n) ________________ is a rational expression that contains fractions in its numerator, denominator, or both.

2. Using Method II to simplify a complex fraction, we multiply both the numerator and the denominator of the complex fraction by the ________________ of all fractions within the complex fraction.

3. Using Method I to simplify a complex fraction, we write the numerator and the denominator each as a single fraction and then multiply the fraction in the numerator by the ________________ of the fraction in the denominator.

4. The expression $\frac{a}{b} \div \frac{c}{d}$ can be written as a(n) ________________, where both the numerator and denominator are single fractions.

Simplifying Complex Fractions

Exercises 1-11: Refer to Examples 1-3 on pages 455-460 in your text and the Section 7.5 lecture video.

Simplify each basic complex fraction.

1. $\dfrac{\frac{2}{3}}{\frac{8}{9}}$ **1.** ______________

2. $\dfrac{2\frac{5}{8}}{3\frac{3}{4}}$ **2.** ______________

3. $\dfrac{\frac{x}{5}}{\frac{2}{5y}}$ **3.** ______________

4. $\dfrac{\frac{(z+2)^2}{3}}{\frac{(z+2)}{9}}$ **4.** ______________

Simplify. Write your answer in lowest terms.

5. $\dfrac{\frac{1}{x}-\frac{1}{y}}{\frac{1}{x}+\frac{1}{y}}$

5. ________________

6. $\dfrac{a+\frac{3}{a}}{a-\frac{3}{a}}$

6. ________________

7. $\dfrac{\frac{3}{x-2}+\frac{1}{x}}{\frac{3}{x}-\frac{1}{x-2}}$

7. ________________

8. $\dfrac{\frac{1}{a}+\frac{1}{b}}{\frac{1}{a^2}-\frac{1}{b^2}}$

8. ________________

Simplify.

9. $\dfrac{\dfrac{5}{x}-\dfrac{3}{x}}{2x}$

9. ________________

10. $\dfrac{\dfrac{1}{x+2}}{\dfrac{4}{x+2}+\dfrac{1}{x}}$

10. ________________

11. $\dfrac{\dfrac{1}{x}-\dfrac{1}{y}}{\dfrac{1}{3x^2}-\dfrac{1}{3y^2}}$

11. ________________

Name: **Course/Section:** **Instructor:**

Chapter 7 Rational Expressions
7.6 Rational Equations and Formulas

Solving Rational Equations ~ Rational Expressions and Equations ~ Graphical and Numerical Solutions ~ Solving a Formula for a Variable ~ Applications

STUDY PLAN

Read: Read Section 7.6 on pages 463-472 in your textbook or eText.

Practice: Do your assigned exercises in your ☐ Book ☐ MyMathLab ☐ Worksheets

Review: Keep your corrected assignments in an organized notebook and use them to review for the test.

Key Terms

Exercises 1-6: Use the vocabulary terms listed below to complete each statement. Note that some terms or expressions may not be used. Some terms may be used more than once.

term
visually
extraneous solution
rational equation
algebraically
greatest common factor
basic rational equation
numerically
least common denominator

1. If an equation contains one or more rational expressions, it is called a(n) ________________.

2. When we solve an equation using the distributive property and the properties of equality, we are solving the equation ________________.

3. To solve a rational equation, do the following.
   Step 1: Find the ________________ of the terms in the equation.
   Step 2: Multiply each side of the equation by the ________________.
   Step 3: Simplify each ________________.
   Step 4: Solve the resulting equation.
   Step 5: Check each answer in the given equation. Any value that makes a denominator equal 0 should be rejected because it is a(n) ________________.

4. When we solve an equation by estimating values from its graph, we are solving the equation ________________.

5. A(n) ________________ has a single rational expression on each side of the equals sign.

6. When we solve an equation using a table of values, we are solving the equation ________________.

Solving Rational Equations

Exercises 1-7: Refer to Examples 1-3 on pages 464-466 in your text and the Section 7.6 lecture video.

Solve each equation.

1. $\frac{4}{3} = \frac{5}{x}$ **1.** ______________

2. $\frac{x-2}{3} = \frac{4x}{3}$ **2.** ______________

3. $\frac{4}{2x-7} = x$ **3.** ______________

4. $\frac{3}{x} - \frac{5}{7} = \frac{2}{x}$ **4.** ______________

Solve each equation. Check your answer.

5. $\frac{1}{a+2} - \frac{1}{a} = \frac{1}{4a}$ **5.** ______________

6. $\frac{1}{x+2}+\frac{1}{x-2}=\frac{5}{x^2-4}$

6. ________________

7. If possible, solve $\frac{2}{x+3}+\frac{1}{x-3}=\frac{6}{x^2-9}$.

7. ________________

Rational Expressions and Equations

Exercises 8-9: Refer to Example 4 on pages 467-468 in your text and the Section 7.6 lecture video.

Determine whether you are given an expression or an equation. If it is an expression, simplify it and then evaluate it for $x=4$. If it is an equation, solve it.

8. $\frac{x^2-3}{x+2}+\frac{1}{x+2}$

8. ________________

____

9. $\frac{x+1}{x-3}=\frac{x}{x-1}$

9. ________________

____

Graphical and Numerical Solutions

Exercise 10: Refer to Example 5 on pages 468-469 in your text and the Section 7.6 lecture video.

10. Solve $\frac{3}{x} = x - 2$ graphically and numerically.

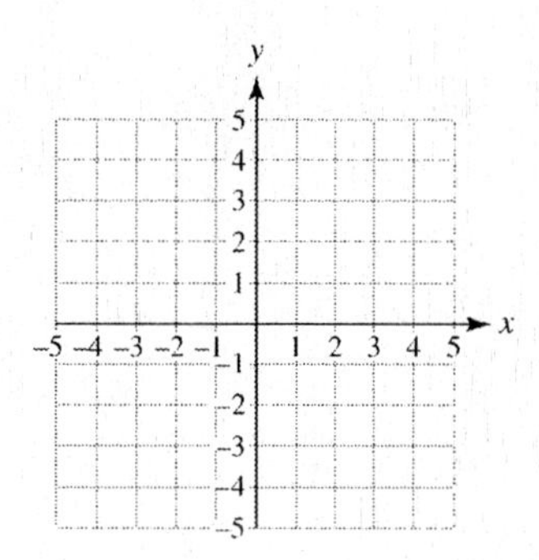

x	−3 −2 −1 0 1 2 3
$y_1 = \frac{3}{x}$	
$y_2 = x - 2$	

Solving a Formula for a Variable

Exercises 11-14: Refer to Examples 6-7 on pages 469-470 in your text and the Section 7.6 lecture video.

11. If a person travels at a speed, or rate, r for time t, then the distance traveled is $d = rt$.

(a) How far does a person travel in 2.5 hours when traveling at 60 miles per hour?

11. (a)______________

(b) Solve the formula $d = rt$ for t.

(b)______________

(c) How long does it take a person to go 250 miles when traveling at 75 miles per hour?

(c)______________

Solve each equation for the specified variable.

12. $A = \frac{bh}{2}$ for h **12.** ______________

13. $P = \frac{nRT}{V}$ for T **13.** ______________

14. $h = \frac{2A}{b_1 + b_2}$ for b_2 **14.** ______________

Applications

Exercises 15-16: Refer to Examples 8-9 on pages 470-472 in your text and the Section 7.6 lecture video.

15. One person can clean a carpet in 2 hours. A second person requires 3 hours to clean the carpet. How long will it take the two people, working together, to clean the carpet? **15.** ______________

16. A runner is competing in a 10-mile race. He runs the first 8 miles at 6 miles per hour, and the last two miles at 7.5 miles per hour. How long does it take him to finish the race? **16.** ______________

Understanding Concepts through Multiple Approaches
(For additional practice, visit MyMathLab.)

17. Solve the equation $\dfrac{2}{x-3}=2.$

(a) Solve algebraically.

(b) Solve numerically using the table shown.

x	-2	0	2	4	6
$y=\dfrac{2}{x-3}$					
$y=2$					

(c) Solve visually.

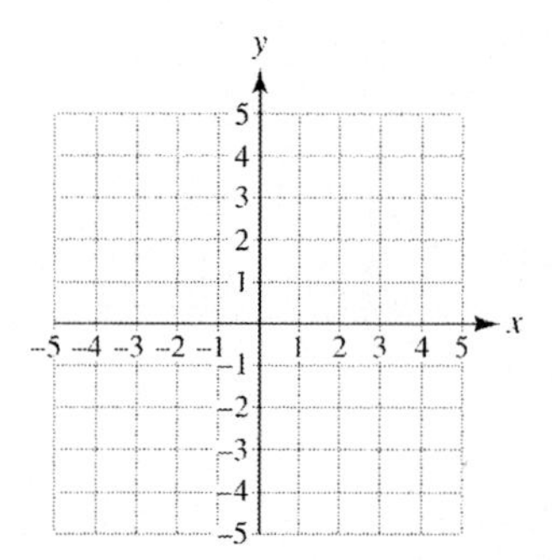

Did you get the same result using each method? Which method do you prefer? Explain why.

Name: ______ **Course/Section:** ______ **Instructor:** ______

Chapter 7 Rational Expressions
7.7 Proportions and Variation

Proportions ~ Direct Variation ~ Inverse Variation ~ Analyzing Data ~ Joint Variation

STUDY PLAN

Read: Read Section 7.7 on pages 476-486 in your textbook or eText.

Practice: Do your assigned exercises in your ☐ Book ☐ MyMathLab ☐ Worksheets

Review: Keep your corrected assignments in an organized notebook and use them to review for the test.

Key Terms

Exercises 1-6: Use the vocabulary terms listed below to complete each statement. Note that some terms or expressions may not be used.

ratio	**directly proportional**
varies inversely	**varies jointly**
proportion	**constant of proportionality**
varies directly	**inversely proportional**
constant of variation	

1. Let x and y denote two quantities. Then y is ________________ to x, or y ________________ with x, if there is a nonzero number k such that $y = \frac{k}{x}$.

2. A(n) ________________ is a comparison of two quantities.

3. Let x and y denote two quantities. Then y is ________________ to x, or y ________________ with x, if there is a nonzero number k such that $y = kx$.

4. A(n) ________________ is a statement that two ratios are equal.

5. Let x, y, and z denote three quantities. Then z ________________ with x and y if there is a nonzero number k such that $z = kxy$.

6. For direct, inverse, or joint variation formulas, the number k is called the ________________, or the ________________.

Proportions

Exercises 1-2: Refer to Examples 1-2 on pages 477-478 in your text and the Section 7.7 lecture video.

1. Eight inches of heavy, wet snow are equivalent to one inch of rain in terms of water content. If 14 inches of this type of snow fall, estimate the water content.

1. ______________

2. A 6-foot-tall person casts a 9-foot-long shadow. If a nearby tree casts a 24-foot-long shadow, estimate the height of the tree.

2. ______________

Direct Variation

Exercises 3-5: Refer to Examples 3-5 on pages 479-481 in your text and the Section 7.7 lecture video.

3. Let y be directly proportional to x, or vary directly with x. Suppose $y = 5$ when $x = 6$. Find y when $x = 8$.

3. ______________

4. The cost of tuition is directly proportional to the number of credits taken. If 8 credits cost $944, find the cost to take14 credits.

4. ______________

5. **(a)** A scatterplot of data is shown in the following figure. Could the data be modeled using a line?

5.(a)

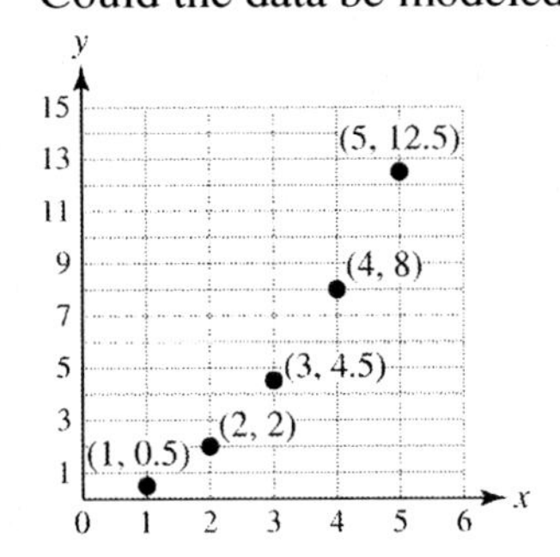

(b) Is y directly proportional to x?

(b)

Inverse Variation

Exercises 6-7: Refer to Examples 6-7 on pages 482-484 in your text and the Section 7.7 lecture video.

6. Let y be inversely proportional to x, or vary inversely with x. Suppose $y = 10$ when $x = 4$. Find y when $x = 8$.

6. ________________

7. **(a)** A scatterplot of data is shown in the following figure. Could the data be modeled using a line?

7.(a)

(b) Is y inversely proportional to x?

(b)________________

(c) Predict the y value if $x = 4$.

(c)________________

Analyzing Data

Exercises 8-10:* *Refer to Example 8 on page 484 in your text and the Section 7.7 lecture video.

Determine whether the data in each table represent direct variation, inverse variation, or neither.

8.

x	3	4	8	12
y	8	6	3	2

8. ______________

9.

x	25	16	9	4
y	5	4	3	2

9. ______________

10.

x	1	3	5	6
y	3	9	15	18

10. ______________

Joint Variation

Exercise 11:* *Refer to Example 9 on page 485 in your text and the Section 7.7 lecture video.

11. The strength S of a rectangular beam varies jointly with its width w and the square of its thickness t. If a beam 5 inches wide and 4 inches thick supports 400 pounds, how much can a similar beam 6 inches wide and 2 inches thick support?

11. ______________

Name: Course/Section: Instructor:

Chapter 8 Radical Expressions
8.1 Introduction to Radical Expressions

Square Roots ~ Cube Roots ~ The Pythagorean Theorem ~ The Distance Formula ~ Graphing (Optional)

STUDY PLAN

Read: Read Section 8.1 on pages 502-509 in your textbook or eText.

Practice: Do your assigned exercises in your ☐ Book ☐ MyMathLab ☐ Worksheets

Review: Keep your corrected assignments in an organized notebook and use them to review for the test.

Key Terms

Exercises 1-8: Use the vocabulary terms listed below to complete each statement. Note that some terms or expressions may not be used.

radicand
distance
radical sign
square root
negative square root
cube root
perfect square
Pythagorean theorem
principal square root

1. The ________________ states that if a right triangle has legs a and b with hypotenuse c, then $a^2 + b^2 = c^2$.

2. The number b is a(n) ________________ of a if $b^3 = a$.

3. The ________________ of 9 is -3 and is denoted $-\sqrt{9}$.

4. If a whole number a has an integer square root, then a is a(n) ________________.

5. In the radical expression $\sqrt{9}$, the symbol $\sqrt{\ }$ is called the ________________ and the number 9 is called the ________________.

6. The number b is a(n) ________________ of a if $b^2 = a$.

7. The ________________ between the points (x_1, y_1) and (x_2, y_2) in the xy-plane is $\sqrt{(x_2 - x_1)^2 + (y_2 - y_1)^2}$.

8. The number 3 is the ________________ of 9 and is denoted $\sqrt{9}$.

Square Roots

Exercises 1-9: Refer to Examples 1-4 on pages 503-505 in your text and the Section 8.1 lecture video.

Find the square roots of each number. Approximate your answer to three decimal places when appropriate.

1. 25 **1.** ____________

2. 10 **2.** ____________

Evaluate each square root.

3. $\sqrt{64}$ **3.** ____________

4. $\sqrt{225}$ **4.** ____________

5. $\sqrt{\frac{4}{9}}$ **5.** ____________

6. When small animals walk, they tend to take fast, short steps, whereas larger animals tend to take slower, longer steps. If an animal is h meter high at the shoulder, then the frequency F in steps per second can be estimated by $F = \frac{0.87}{\sqrt{h}}$. The value of F is referred to as the animal's stepping frequency.

(a) Estimate the stepping frequency for a dog 0.6 meters at the shoulder **6. (a)**____________

(b) Estimate the stepping frequency for a giraffe 4 meters at the shoulder. **(b)**____________

State whether each expression is one or more of the following: a real number, a rational number, an irrational number, or none.

7. $\sqrt{169}$ **7.** ______________

8. $\sqrt{-25}$ **8.** ______________

9. $\sqrt{8}$ **9.** ______________

Cube Roots

Exercises 10-12: Refer to Example 5 on page 505 in your text and the Section 8.1 lecture video.

Find the cube root of each expression. Approximate the result to three decimal places when appropriate.

10. $\sqrt[3]{64}$ **10.** ______________

11. $-\sqrt[3]{216}$ **11.** ______________

12. $\sqrt[3]{18}$ **12.** ______________

The Pythagorean Theorem

Exercises 13-15: Refer to Examples 6-7 on pages 506-507 in your text and the Section 8.1 lecture video.

A right triangle has legs a and b and hypotenuse c. Find the length of the missing side.

13. $a = 3$ inches, $b = 4$ inches **13.** ______________

14. $b = 8$ inches, $c = 10$ inches **14.** ______________

15. The bases in a major league baseball field are placed at the four corners of a square 90 feet on a side. Approximate, to the nearest foot, the distance d from the center of the infield to the batters' box.

15. ______________

The Distance Formula

Exercises 16-17: Refer to Examples 8-9 on page 508 in your text and the Section 8.1 lecture video.

16. Find the exact length of the line segment between the points $(-3,7)$ and $(1,3)$. Then approximate this value to two decimal places.

16. ______________

17. Two cars stop at an intersection. The first car travels south at 40 miles per hour and the second car travels west at 30 miles per hour. Find the distance between the cars after 45 minutes.

17. ______________

Graphing (Optional)

Exercise 18: Refer to Example 10 on page 509 in your text and the Section 8.1 lecture video.

18. Make a table of values for $y=\sqrt{x-3}$. Then sketch the graph.

18. ______________

x	$\sqrt{x-3}$
3	
4	
7	
12	

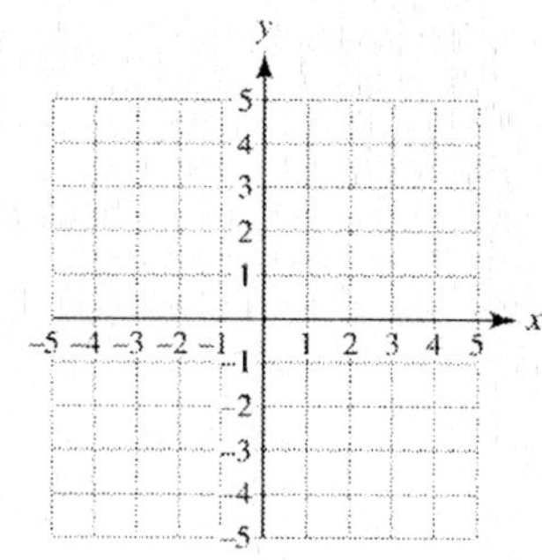

Name: Course/Section: Instructor:

Chapter 8 Radical Expressions
8.2 Multiplication and Division of Radical Expressions

The Product Rule ~ Simplifying Square Roots ~ The Quotient Rule

STUDY PLAN

Read: Read Section 8.2 on pages 513-518 in your textbook or eText.

Practice: Do your assigned exercises in your ☐ Book ☐ MyMathLab ☐ Worksheets

Review: Keep your corrected assignments in an organized notebook and use them to review for the test.

The Product Rule

Exercises 1-7: Refer to Examples 1-3 on pages 514-515 in your text and the Section 8.2 lecture video.

Multiply each expression.

1. $\sqrt{3}\cdot\sqrt{12}$ **1.** ______________

2. $\sqrt{3}\cdot\sqrt{27}$ **2.** ______________

Multiply each pair of radicals. Assume that all variables are positive.

3. $\sqrt{7}\cdot\sqrt{2ab}$ **3.** ______________

4. $\sqrt{5x}\cdot\sqrt{3y}$ **4.** ______________

Write each square root as a product of square roots.

5. $\sqrt{14}$ **5.** ______________

6. $\sqrt{5a};\ a>0$ **6.** ______________

7. $\sqrt{33}$ **7.** ______________

Simplifying Square Roots

Exercises 8-17: Refer to Examples 4-6 on pages 515-516 in your text and the Section 8.2 lecture video.

Simplify each expression.

8. $\sqrt{28}$ **8.** ______________

9. $\sqrt{75}$ **9.** ______________

10. $\sqrt{242}$ **10.** ______________

Simplify each expression. Assume that all variables are positive.

11. $\sqrt{x^6}$ **11.** ______________

12. $\sqrt{a^5}$ **12.** ______________

13. $\sqrt{r^2t^4}$ **13.** ______________

14. $\sqrt{36z^2}$ **14.** ______________

15. $\sqrt{49x^3}$ **15.** ______________

16. $\sqrt{25x^6y^5}$ **16.** ______________

17. $\sqrt{2a}\cdot\sqrt{50a^3}$ **17.** ______________

The Quotient Rule

Exercises 18-23: Refer to Examples 7-8 on pages 517-518 in your text and the Section 8.2 lecture video.

Simplify each expression. Assume that all variables are positive.

18. $\sqrt{\frac{25}{49}}$ **18.** ____________

19. $\sqrt{\frac{9}{z^6}}$ **19.** ____________

20. $\sqrt{\frac{7}{64x^2}}$ **20.** ____________

21. $\frac{\sqrt{48}}{\sqrt{16}}$ **21.** ____________

22. $\frac{\sqrt{50xy}}{\sqrt{2xy}}$ **22.** ____________

23. $\sqrt{6ab^2} \cdot \frac{\sqrt{6}}{\sqrt{a^3b^4}}$ **23.** ____________

Name: Course/Section: Instructor:

Chapter 8 Radical Expressions
8.3 Addition and Subtraction of Radical Expressions

Addition of Radical Expressions ~ Subtraction of Radical Expressions

STUDY PLAN

Read: Read Section 8.3 on pages 522-526 in your textbook or eText.

Practice: Do your assigned exercises in your ☐ Book ☐ MyMathLab ☐ Worksheets

Review: Keep your corrected assignments in an organized notebook and use them to review for the test.

Addition of Radical Expressions

Exercises 1-13: Refer to Examples 1-6 on pages 522-525 in your text and the Section 8.3 lecture video.

Write each pair of terms as like radicals, if possible.

1. $\sqrt{32}, \sqrt{50}$ **1.** ____________

2. $\sqrt{108}, \sqrt{18}$ **2.** ____________

Simplify each expression.

3. $4\sqrt{3}+5\sqrt{3}$ **3.** ____________

4. $3\sqrt[3]{10}+4\sqrt[3]{10}$ **4.** ____________

5. $4\sqrt{32}+2\sqrt{50}$ **5.** ____________

6. $\sqrt{5}+\sqrt{45}+\sqrt{80}$ **6.** ____________

Add each expression and simplify. Assume that all variables are positive.

7. $\sqrt{a}+5\sqrt{a}$

7. ______________

8. $2\sqrt[3]{rt}+8\sqrt[3]{rt}$

8. ______________

9. $3\sqrt{16x}+\sqrt{25x}$

9. ______________

Simplify each expression. Assume that variables are positive.

10. $2\sqrt{12}+\sqrt{27}+4\sqrt{3}$

10. ______________

11. $5\sqrt{x^3}+x\sqrt{x}+\sqrt{16x^2}$

11. ______________

12. Find the exact perimeter of a rectangle with length $\sqrt{54}$ inches and width $\sqrt{24}$ inches.

12. ______________

13. Suppose that two open flood channels have flow rates R_1 and R_2 given by $R_1 = 1000\sqrt{m_1}$ and $R_2 = 1500\sqrt{m_2}$, where R_1 and R_2 are in cubic feet per second and m_1 and m_2 are the slopes of the channels, respectively.

(a) Find $R_1 + R_2$ if $m_1 = 0.16$ and $m_2 = 0.09$.

13. (a) ______________

(b) Find $R_1 + R_2$ if both channels have slope m.

(b) ______________

Subtraction of Radical Expressions

Exercises 14-19: Refer to Examples 7-8 on pages 525-526 in your text and the Section 8.3 lecture video.

Subtract each expression and simplify. Assume that the variable x is positive.

14. $7\sqrt{2}-2\sqrt{2}$

14. ______________

15. $5\sqrt{20}-2\sqrt{45}$

15. ______________

16. $\sqrt{25x}-3\sqrt{x}$

16. ______________

Subtract each expression and simplify. Assume that all variables are positive.

17. $9\sqrt{a^3}-\sqrt{4a}$

17. ______________

18. $8\sqrt{xy^2}-2\sqrt{xy^2}$

18. ______________

19. $10\sqrt[3]{x}-2\sqrt[3]{x}$

19. ______________

Name: ______ **Course/Section:** ______ **Instructor:** ______

Chapter 8 Radical Expressions
8.4 Simplifying Radical Expressions

Simplifying Products ~ Rationalizing the Denominator

STUDY PLAN

Read: Read Section 8.4 on pages 528-531 in your textbook or eText.

Practice: Do your assigned exercises in your ☐ Book ☐ MyMathLab ☐ Worksheets

Review: Keep your corrected assignments in an organized notebook and use them to review for the test.

Key Terms

Exercises 1-3: Use the vocabulary terms listed below to complete each statement. Note that some terms or expressions may not be used.

conjugate
like radicals
rationalizing the denominator

1. Square roots with the same radicand are ________________, as are cube roots with the same radicand.

2. One way to standardize quotients containing radical expressions is to remove the radical expressions from the denominator by a process called ________________.

3. The ________________ of $\sqrt{a} - \sqrt{b}$ is $\sqrt{a} + \sqrt{b}$.

Simplifying Products

Exercises 1-2: Refer to Example 1 on page 528 in your text and the Section 8.4 lecture video.

Multiply and simplify.

1. $(\sqrt{5}+2)(\sqrt{5}-2)$ **1.** ______________

2. $(\sqrt{x}-4)(\sqrt{x}+7)$; $x \geq 0$ **2.** ______________

Rationalizing the Denominator

Exercises 3-11: Refer to Examples 2-4 on pages 529-530 in your text and the Section 8.4 lecture video.

Rationalize the denominator of each expression. Assume that all variables are positive.

3. $\frac{2}{\sqrt{5}}$ **3.** ______________

4. $\frac{7}{3\sqrt{7}}$ **4.** ______________

5. $-\frac{2}{\sqrt{x}}$ **5.** ______________

6. $\sqrt{\frac{5}{y}}$ **6.** ______________

Write the conjugate of each expression.

7. $2+\sqrt{3}$ **7.** ______________

8. $\sqrt{5}-3$ **8.** ______________

9. $\sqrt{a}-\sqrt{b}$ **9.** ______________

Rationalize the denominator of each expression.

10. $\frac{2-\sqrt{3}}{2+\sqrt{3}}$ **10.** ______________

11. $\frac{\sqrt{a}+\sqrt{b}}{\sqrt{a}-\sqrt{b}}$ **11.** ______________

Name: Course/Section: Instructor:

Chapter 8 Radical Expressions
8.5 Equations Involving Radical Expressions

Solving Radical Equations ~ Solving an Equation for a Variable ~ Numerical and Graphical Methods (Optional)

STUDY PLAN

Read: Read Section 8.5 on pages 533-539 in your textbook or eText.

Practice: Do your assigned exercises in your ☐ Book ☐ MyMathLab ☐ Worksheets

Review: Keep your corrected assignments in an organized notebook and use them to review for the test.

Key Terms

Exercises 1-3: Use the vocabulary terms listed below to complete each statement. Note that some terms or expressions may not be used.

radical equation
extraneous solution
squaring property for solving equations

1. An equation that has a variable as a radicand is an example of a(n) ________________.

2. The ________________ states that if each side of an equation is squared, then any solutions to the given equation are among the solutions to the new equation.

3. An example of a(n) ________________ is a value that satisfies a new equation generated by squaring both sides of an equation, but does not satisfy the given equation.

Solving Radical Equations

Exercises 1-5: Refer to Examples 1-5 on pages 534-537 in your text and the Section 8.5 lecture video.

1. Solve the equation $30 = 1.22\sqrt{h}$ to find the elevation in feet necessary for a person to see 30 miles to the horizon. Round to the nearest foot.

1. ______________

2. Solve $\sqrt{x+5} = 6$.

2. ______________

3. Solve $\sqrt{x+6} = x$.

3. ______________

4. Solve $\sqrt{3-a} + 2 = 9$.

4. ______________

5. Solve $3\sqrt{t} + 4 = t$.

5. ______________

Solving an Equation for a Variable

Exercises 6-7: Refer to Examples 6-7 on pages 537-538 in your text and the Section 8.5 lecture video.

6. If a circular highway curve without any banking has a radius of r feet, then the speed limit S in miles per hour for the curve can be estimated by the equation $S = 1.5\sqrt{r}$.

(a) Use the formula to find the speed limit for a curve with a 400-foot radius.

6. (a)______________

(b) Use the formula to find a safe radius for a curve with a 45-mile-per-hour speed limit.

(b)______________

7. The period P in seconds of a pendulum can be estimated with the formula $P = 2\pi\sqrt{\frac{L}{32}}$, where L is the length of the pendulum in feet.

(a) If the pendulum is 3 feet long, find its period. Round to the nearest foot.

7. (a)______________

(b) Find L if the period is 2.5 seconds. Round to the nearest foot.

(b)______________

Understanding Concepts through Multiple Approaches
(For additional practice, visit MyMathLab.)

8. Solve the equation $2\sqrt{x+3}=12$.

(a) Solve algebraically.

(b) Solve numerically using the table feature of a graphing calculator.

(c) Solve visually using a graphing calculator.

Did you get the same result using each method? Which method do you prefer? Explain why.

Name: ______ Course/Section: ______ Instructor: ______

Chapter 8 Radical Expressions
8.6 Higher Roots and Rational Exponents

Higher Roots ~ Rational Exponents

STUDY PLAN

Read: Read Section 8.6 on pages 543-548 in your textbook or eText.

Practice: Do your assigned exercises in your ☐ Book ☐ MyMathLab ☐ Worksheets

Review: Keep your corrected assignments in an organized notebook and use them to review for the test.

Key Terms

Exercises 1-6: Use the vocabulary terms or expressions listed below to complete each statement. Note that some terms or expressions may not be used.

index	$\sqrt[n]{a}$	$\frac{\sqrt[n]{a}}{\sqrt[n]{b}}$
***n*th root**	$\frac{1}{a^{m/n}}$	$\sqrt[n]{ab}$
odd root		
even root		
principal *n*th root	$\left(\sqrt[n]{a}\right)^m$	$\sqrt[n]{a^m}$

1. The number b is a(n) ________ of a if $b^n = a$, where n is a positive integer. An odd n signifies a(n) ________, and an even n signifies a(n) ________.

2. The ________ of a is denoted $\sqrt[n]{a}$, where n is called the ________.

3. Let a and b be real numbers, where $\sqrt[n]{a}$ and $\sqrt[n]{b}$ are both defined. Then $\sqrt[n]{a} \cdot \sqrt[n]{b} =$ ________ and $\sqrt[n]{\frac{a}{b}} =$ ________ with $b \neq 0$.

4. If n is an integer greater than 1, then $a^{1/n} =$ ________. If $a < 0$ and n is an even positive integer, then $a^{1/n}$ is not a real number.

5. If m and n are positive integers with $\frac{m}{n}$ in lowest terms, then $a^{m/n} =$ ________ or $a^{m/n} =$ ________. If $a < 0$ and n is an even positive integer, then $a^{m/n}$ is not a real number.

6. If m and n are positive integers with $\frac{m}{n}$ in lowest terms, then $a^{-m/n} =$ ________, $a \neq 0$.

Higher Roots

Exercises 1-12: Refer to Examples 1-3 on pages 543-544 in your text and the Section 8.5 lecture video.

Find each root, if possible.

1. $\sqrt[3]{-64}$ **1.** ________

2. $\sqrt[4]{16}$ **2.** ________

3. $\sqrt[4]{-81}$ **3.** ________

4. $-\sqrt[5]{243}$ **4.** ________

Simplify each product.

5. $\sqrt[3]{-5} \cdot \sqrt[3]{25}$ **5.** ________

6. $\sqrt[4]{9} \cdot \sqrt[4]{9}$ **6.** ________

7. $\sqrt[5]{64} \cdot \sqrt[5]{-16}$ **7.** ________

8. $\sqrt[6]{\frac{3}{4}} \cdot \sqrt[6]{\frac{4}{3}}$ **8.** ________

Simplify each quotient.

9. $\sqrt[3]{\frac{8}{27}}$ **9.** ____________

10. $\sqrt[4]{\frac{8}{625}}$ **10.** ____________

11. $\frac{\sqrt[3]{-81}}{\sqrt[3]{3}}$ **11.** ____________

12. $\frac{\sqrt[5]{160}}{\sqrt[5]{5}}$ **12.** ____________

Rational Exponents

Exercises 13-22: Refer to Examples 4-7 on pages 545-547 in your text and the Section 8.5 lecture video.

Write each expression in radical notation and then evaluate.

13. $9^{1/2}$ **13.** ____________

14. $81^{1/4}$ **14.** ____________

15. $125^{1/3}$ **15.** ____________

16. $-81^{3/4}$

16. ______________

17. $(-64)^{2/3}$

17. ______________

18. $32^{3/5}$

18. ______________

19. Biologists have found that the weight W in kilograms of some types of birds and the length L in meters of their wing span are related by the formula $L = 0.91W^{1/3}$.

(a) Use radical notation to write this formula.

19. (a)______________

(b) Estimate the wing span of a 1.5-kilogram bird to two decimal places.

(b)______________

Write each expression in radical notation and then evaluate.

20. $243^{-3/5}$

20. ______________

21. $-16^{-3/4}$

21. ______________

22. $(-125)^{-2/3}$

22. ______________

Name: **Course/Section:** **Instructor:**

Chapter 9 Quadratic Equations
9.1 Parabolas

Graphing Parabolas ~ The Graph of $y = ax^2$

STUDY PLAN

Read: Read Section 9.1 on pages 559-565 in your textbook or eText.

Practice: Do your assigned exercises in your ☐ Book ☐ MyMathLab ☐ Worksheets

Review: Keep your corrected assignments in an organized notebook and use them to review for the test.

Key Terms

Exercises 1-5: Use the vocabulary terms listed below to complete each statement. Note that some terms or expressions may not be used. Some terms may be used more than once.

wider	**highest**
vertex	**narrower**
lowest	***x*-coordinate**
upward	***y*-coordinate**
parabola	**axis of symmetry**
downward	

1. The equation of a ________________ with a vertical axis of symmetry can be written $y = ax^2 + bx + c$, where *a*, *b*, and *c* are constants with $a \neq 0$. If $a > 0$, the parabola opens ________________. If $a < 0$, the parabola opens ________________.

2. The ________________ is the ________________ point on the graph of a parabola that opens upward or the ________________ point on the graph of a parabola that opens downward.

3. The ________________ is a line about which a parabola is symmetric.

4. The ________________ of the vertex of the graph of $y = ax^2 + bx + c,\ a \neq 0,$ is given by $x = -\dfrac{b}{2a}$.

5. The graph of $y = ax^2$ is a parabola with the following characteristics.
   The ________________ is $(0,0)$ and the ________________ is given by $x = 0$.
   It opens ________________ if $a > 0$ and opens ________________ if $a < 0$.
   It is ________________ than the graph of $y = x^2$ if $0 < |a| < 1$.
   It is ________________ than the graph of $y = x^2$ if $|a| > 1$.

Graphing Parabolas

Exercises 1-9: Refer to Examples 1-5 on pages 560-563 in your text and the Section 9.1 lecture video.

Determine whether each parabola opens upward or downward.

1. $y = 2x^2 + 3x + 1$ **1.** ______________

2. $y = 8 + 4x - 5x^2$ **2.** ______________

3. $y = -(x+3)^2$ **3.** ______________

4. Find the vertex and axis of symmetry for the graph of $y = -x^2 + 6x - 5$. **4.** ______________

____

5. Graph $y = 2x^2 + 8x + 8$.

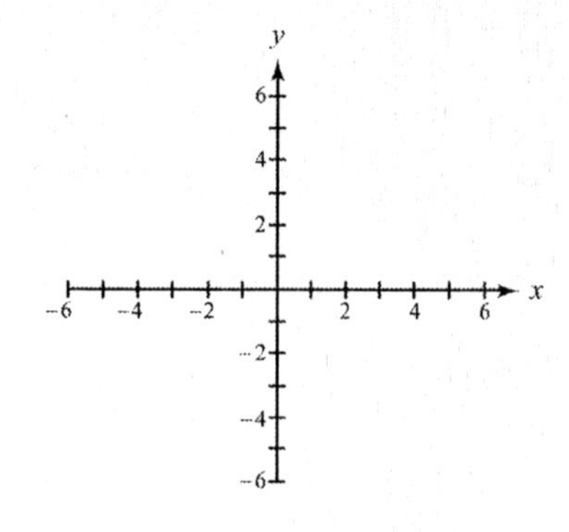

Graph each equation. Identify the vertex and the axis of symmetry.

6. $y = -x^2 - 2x + 3$

6. ________________

____

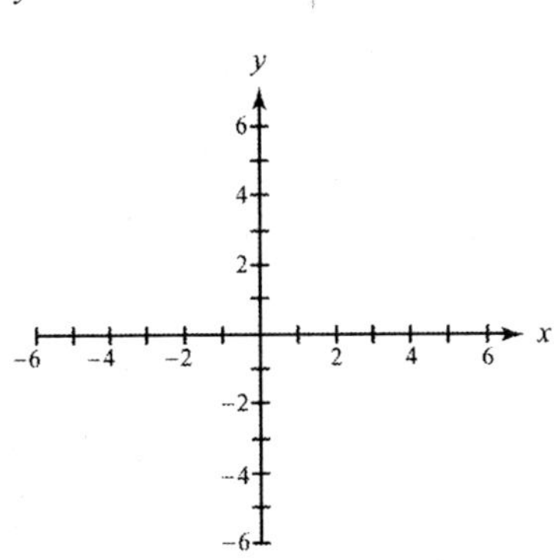

7. $y = 4x - x^2$

7. ________________

____

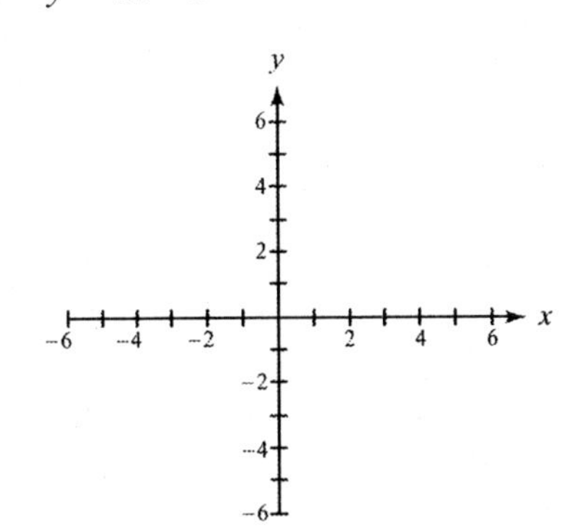

8. $y = (x-3)^2$

8. ________________

____

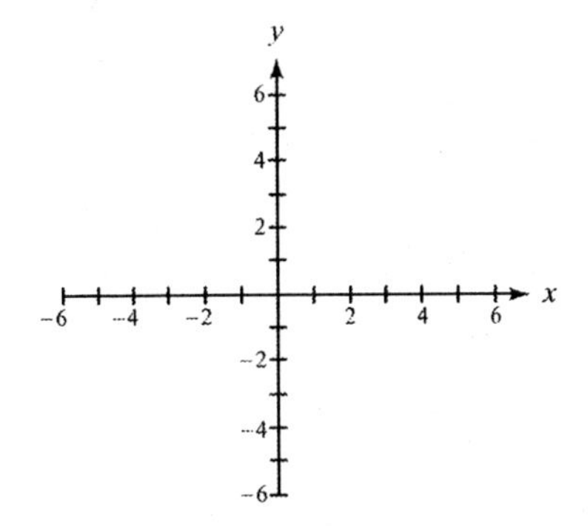

9. When a golf ball is hit into the air, its height y in feet after x seconds can be calculated by the equation $y = -16x^2 + 112x$.

(a) Find the vertex of this parabola.

9. (a)________________

(b) Interpret the x- and y-coordinates of the vertex.

(b)________________

____

The Graph of $y = ax^2$

Exercise 10: Refer to Example 6 on page 564 in your text and the Section 9.1 lecture video.

10. Compare the graph of $y = \frac{1}{2}x^2$ to the graph of $y = x^2$. **10.** ____________

Then graph both functions on the same coordinate axes.

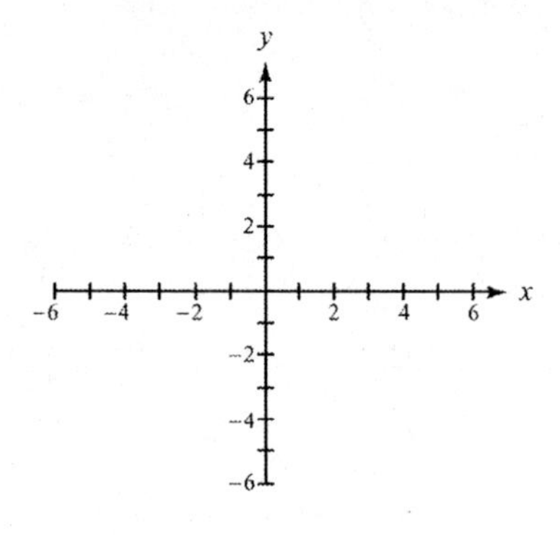

Name: **Course/Section:** **Instructor:**

Chapter 9 Quadratic Equations
9.2 Introduction to Quadratic Equations

Basics of Quadratic Equations ~ Symbolic, Graphical, and Numerical Solutions ~ The Square Root Property

STUDY PLAN

Read: Read Section 9.2 on pages 567-572 in your textbook or eText.

Practice: Do your assigned exercises in your ☐ Book ☐ MyMathLab ☐ Worksheets

Review: Keep your corrected assignments in an organized notebook and use them to review for the test.

Key Terms

Exercises 1-2: Use the vocabulary terms listed below to complete each statement. Note that some terms or expressions may not be used.

quadratic equation
square root property

1. A(n) ________________ is an equation that can be written as $ax^2 + bx + c = 0$, where a, b, and c are constants with $a \neq 0$.

2. The ________________ states that if k is a nonnegative number, then the solutions to the equation $x^2 = k$ are given by $x = \pm\sqrt{k}$. If $k < 0$, then this equation has no real solutions.

Symbolic, Graphical, and Numerical Solutions

***Exercises 1-3:** Refer to Example 1 on pages 569-570 in your text and the Section 9.2 lecture video.*

Solve each quadratic equation. Support your results numerically and graphically.

1. $3x^2 + 1 = 0$ **1.** ____________

2. $x^2 + 5x + 6 = 0$ **2.** ____________

3. $x^2 - 3x = 4$ **3.** ____________

The Square Root Property

***Exercises 4-8:** Refer to Examples 2-4 on pages 570-572 in your text and the Section 9.2 lecture video.*

Solve each equation.

4. $x^2 = 5$ **4.** ____________

5. $25x^2 - 16 = 0$ **5.** ____________

6. $(x-3)^2 = 36$ **6.** ____________

7. An object falls from a height of 72 feet. Use the formula $d = h - 16t^2$ to determine how long it takes for the object to hit the ground. Round to the nearest hundredth of a second. **7.** ____________

8. Solve the equation $15x^2 + 30 = 265$. **8.** ____________

Name: ______ Course/Section: ______ Instructor: ______

Chapter 9 Quadratic Equations
9.3 Solving by Completing the Square

Perfect Square Trinomials ~ Completing the Square

STUDY PLAN

Read: Read Section 9.3 on pages 575-578 in your textbook or eText.

Practice: Do your assigned exercises in your ☐ Book ☐ MyMathLab ☐ Worksheets

Review: Keep your corrected assignments in an organized notebook and use them to review for the test.

Perfect Square Trinomials

Exercises 1-3: Refer to Example 1 on pages 575-576 in your text and the Section 9.3 lecture video.

Determine whether each trinomial is a perfect square trinomial.

1. $x^2 - 6x + 9$ **1.** ________

2. $x^2 + 9x + 81$ **2.** ________

3. $x^2 + 12x + 36$ **3.** ________

Completing the Square

Exercises 4-7: Refer to Examples 2-5 on pages 576-578 in your text and the Section 9.3 lecture video.

4. Find the term that should be added to $x^2 + 14x$ to form a perfect square trinomial. Write the resulting perfect square trinomial in factored form.

4. ________________

5. Solve the equation $x^2 - 8x + 2 = 0$.

5. ________________

6. Solve the equation $x^2 + 7x = 5$.

6. ________________

7. Solve the equation $3x^2 + 2x = 4$.

7. ________________

Name: **Course/Section:** **Instructor:**

Chapter 9 Quadratic Equations
9.4 The Quadratic Formula

Solving Quadratic Equations ~ The Discriminant

STUDY PLAN

Read: Read Section 9.4 on pages 580-585 in your textbook or eText.

Practice: Do your assigned exercises in your ☐ Book ☐ MyMathLab ☐ Worksheets

Review: Keep your corrected assignments in an organized notebook and use them to review for the test.

Key Terms

Exercises 1-3: Use the vocabulary terms listed below to complete each statement. Note that some terms or expressions may not be used.

no
one
two
discriminant
quadratic formula

1. The solutions to $ax^2+bx+c=0$ with $a\neq 0$ are given by $x=\dfrac{-b\pm\sqrt{b^2-4ac}}{2a}$.

This is called the ____________________.

2. The expression b^2-4ac in the quadratic formula is called the ____________________.

3. If $b^2-4ac>0$, there is/are ____________________ real solution(s).

If $b^2-4ac=0$, there is/are ____________________ real solution(s).

If $b^2-4ac<0$, there is/are ____________________ real solution(s).

Solving Quadratic Equations

Exercises 1-8: Refer to Examples 1-6 on pages 581-584 in your text and the Section 9.4 lecture video.

Determine a, b, and c by writing each equation in the form $ax^2 + bx + c = 0$.

1. $2x^2 + 5x - 1 = 0$ **1.** ____________

2. $-3x^2 + 4 = 5x$ **2.** ____________

3. $(x+4)(x-4) = 0$ **3.** ____________

4. Solve $2x^2 - 7x - 15 = 0$ by factoring and by the quadratic formula. **4.** ____________

Solve each equation. Support your result graphically.

5. $3x^2 + x = 1$ **5.** ____________

6. $x^2 - 6x + 9 = 0$ **6.** ____________

7. $-4x^2 + 3x - 5 = 0$ **7.** ____________

8. Solve the equation $x^2 + x + 2 = 3x^2 - 2$. **8.** ____________

The Discriminant

Exercises 9-10: Refer to Examples 7-8 on pages 584-585 in your text and the Section 9.4 lecture video.

9. Use the discriminant to determine the number of solutions to $9x^2 = 25$.

9. ________________

10. A graph of $y = ax^2 + bx + c$ is shown.

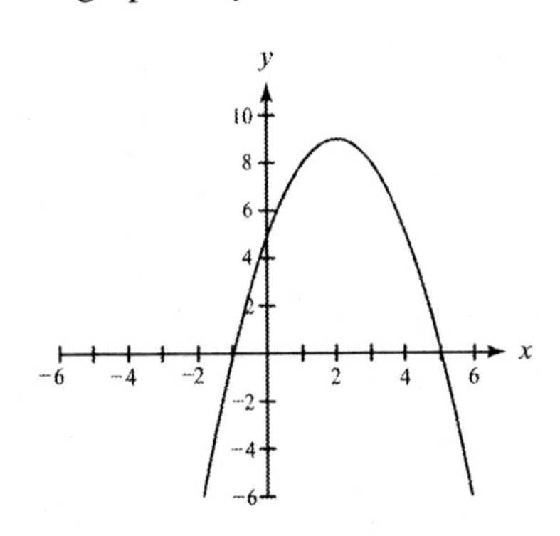

(a) State whether $a > 0$ or $a < 0$.

(b) Solve the equation $ax^2 + bx + c = 0$.

(c) Determine whether the discriminant is positive, negative, or zero.

10. (a)______________

(b)______________

(c)______________

Name: Course/Section: Instructor:

Chapter 9 Quadratic Equations
9.5 Complex Solutions

Basic Concepts ~ Addition, Subtraction, and Multiplication ~ Quadratic Equations with Complex Solutions

STUDY PLAN

Read: Read Section 9.5 on pages 588-594 in your textbook or eText.

Practice: Do your assigned exercises in your ☐ Book ☐ MyMathLab ☐ Worksheets

Review: Keep your corrected assignments in an organized notebook and use them to review for the test.

Key Terms

Exercises 1-3: Use the vocabulary terms listed below to complete each statement. Note that some terms or expressions may not be used.

real part	**standard form**
imaginary unit	**imaginary part**
complex number	**imaginary number**

1. The number $i = \sqrt{-1}$, where $i^2 = -1$, is called the ________________.

2. A(n) ________________ can be written in ________________ $a + bi$, where a and b are real numbers. The ________________ is a and the ________________ is b.

3. A complex number $a + bi$ with $b \neq 0$ is a(n) ________________.

Basic Concepts

Exercises 1-3: Refer to Example 1 on page 590 in your text and the Section 9.5 lecture video.

Write each square root using the imaginary unit i.

1. $\sqrt{-9}$ **1.** ______________

2. $\sqrt{-5}$ **2.** ______________

3. $\sqrt{-12}$ **3.** ______________

Addition, Subtraction, and Multiplication

Exercises 4-7: Refer to Examples 2-3 on pages 590-591 in your text and the Section 9.5 lecture video.

Write each sum or difference in standard form.

4. $(-4+i)-(6-8i)$ **4.** ______________

5. $3i+(7-2i)$ **5.** ______________

Write each product in standard form.

6. $(-2+3i)(3-i)$ **6.** ______________

7. $(3+4i)^2$ **7.** ______________

Quadratic Equations with Complex Solutions

Exercises 8-10: Refer to Examples 4-6 on pages 592-593 in your text and the Section 9.5 lecture video.

8. Solve $x^2 + 8 = 0$. **8.** ________________

9. Solve $x^2 - 3x + 5 = 0$. Write your answer in standard form: $a + bi$ **9.** ________________

10. Solve $\frac{1}{3}x^2 + 2 = x$. Write your answer in standard form: $a + bi$ **10.** ________________

Name: ______________ **Course/Section:** ______________ **Instructor:** ______________

Chapter 9 Quadratic Equations
9.6 Introduction to Functions

Basic Concepts ~ Representing a Function ~ Definition of a Function ~ Identifying a Function

STUDY PLAN

Read: Read Section 9.6 on pages 595-603 in your textbook or eText.

Practice: Do your assigned exercises in your ☐ Book ☐ MyMathLab ☐ Worksheets

Review: Keep your corrected assignments in an organized notebook and use them to review for the test.

Key Terms

Exercises 1-4: Use the vocabulary terms listed below to complete each statement. Note that some terms or expressions may not be used.

input
range
diagram
function
domain
output
function notation
vertical line test

1. A(n) ________________ is a set of ordered pairs (x, y), where each x-value corresponds to exactly one y-value.

2. The ________________ of f is the set of all x-values, and the ________________ of f is the set of all y-values.

3. The ________________ states that if every vertical line intersects a graph at no more than one point, then the graph represents a ________________.

4. The notation $y = f(x)$ is called ________________. The ________________ is x, the ________________ is y, and the name of the function is f.

Representing a Function

Exercises 1-5: Refer to Examples 1-3 on pages 598-600 in your text and the Section 9.6 lecture video.

Evaluate $f(x)$ at the given value of x.

1. $f(x) = 2x - 9,\ x = -3$ **1.** ______________

2. $f(x) = \dfrac{-3x}{x-1},\ x = 4$ **2.** ______________

3. $f(x) = \sqrt{6x-2},\ x = 3$ **3.** ______________

4. Sketch a graph of $f(x) = -x - 2$. Use the graph to evaluate $f(1)$. **4.** 

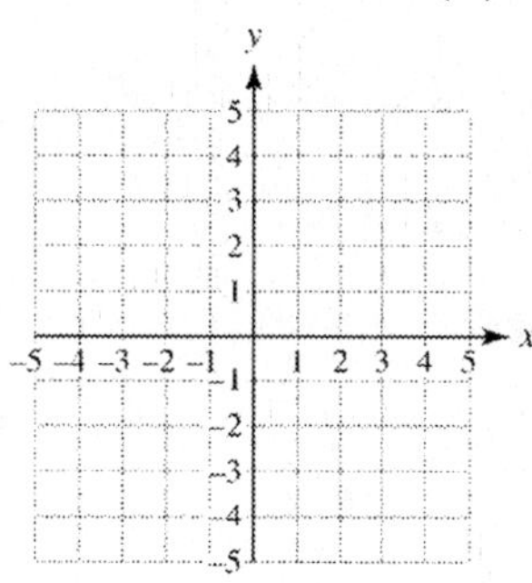

5. Let a function f square the input x and then subtract 4 to obtain the output y.

(a) Write a formula for $f(x)$. **5. (a)**________________

(b) Make a table of values for f. Use $x = -2, -1, 0, 1, 2$. **(b)**________________

x	-2	-1	0	1	2
$f(x)$					

(c) Sketch a graph of $y = f(x)$.

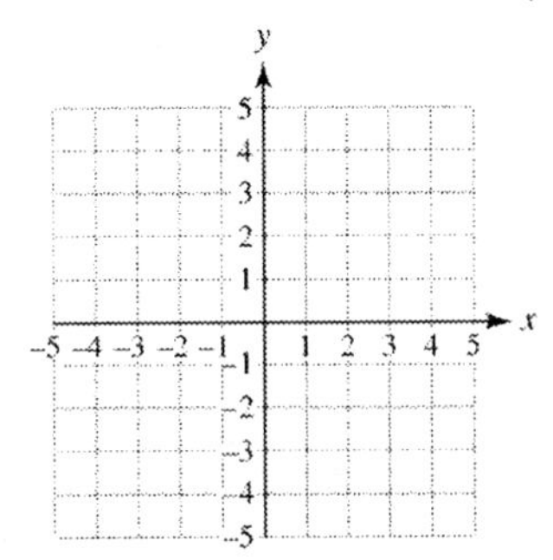

Definition of a Function

Exercises 6-8: Refer to Example 4 on pages 600-601 in your text and the Section 9.6 lecture video.

Use $f(x)$ to find the domain of f.

6. $f(x) = 2x + 5$ **6.**________________

7. $f(x) = \dfrac{x-3}{x}$ **7.**________________

8. $f(x) = \sqrt{x-4}$ **8.**________________

Identifying a Function

Exercises 9-11: Refer to Examples 5-6 on pages 601-602 in your text and the Section 9.6 lecture video.

9. Determine whether the table represents a function.

9. 

x	3	0	−4	5	−2
y	1	−2	4	2	4

Determine whether the graphs shown represent functions.

10.

10. 

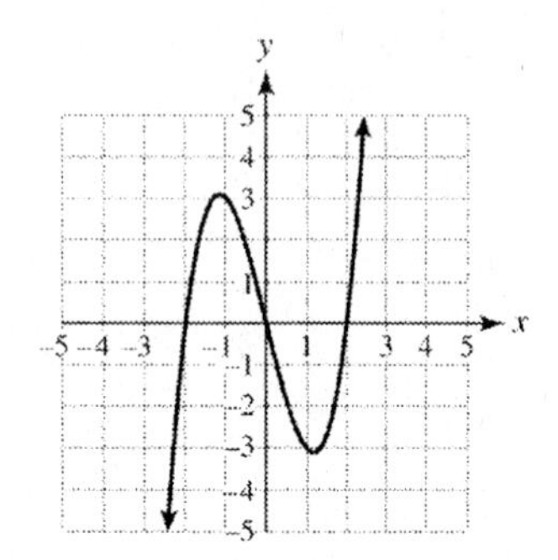

11.

11. ________________

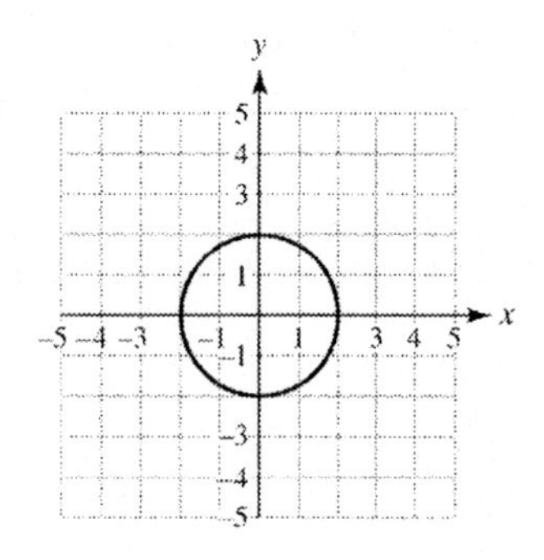

Chapter 1 Introduction to Algebra

1.1 Numbers, Variables, and Expressions

Key Terms

1. formula
2. whole numbers
3. prime number
4. product; factors
5. prime factorization
6. variable
7. natural numbers
8. algebraic expression
9. composite number
10. equation

Prime Numbers and Composite Numbers

1. prime
2. neither
3. composite; $65 = 5 \cdot 13$
4. composite; $180 = 2 \cdot 2 \cdot 3 \cdot 3 \cdot 5$

Variables, Algebraic Expressions, and Equations

5. 10
6. 12
7. 15
8. 2
9. 20
10. -6
11. 3
12. 15
13. 1
14. 24

Translating Words to Expressions

15. $n - 5$; n is the number
16. $3d$; d is the cost of a DVD
17. $y(x-5)$; x is one number, y is the other number
18. $\frac{100}{n}$; n is the number
19. **(a)** \$1.50
    (b) $P = 0.15t$
    (c) \$3.00
20. **(a)** $V = lwh$
    (b) 180 cubic inches

1.2 Fractions

Key Terms

1. lowest terms
2. least common denominator (LCD)
3. basic principle of fractions
4. multiplicative inverse; reciprocal
5. greatest common factor (GCF)

Basic Concepts

1. numerator: 7
denominator: 15

2. numerator: x
denominator: yz

3. numerator: $a+2$
denominator: $b-3$

Simplifying Fractions to Lowest Terms

4. 14

5. 8

6. $\frac{2}{5}$

7. $\frac{3}{7}$

Multiplication and Division of Fractions

8. $\frac{12}{35}$

9. $\frac{4}{7}$

10. $\frac{5}{2}$

11. $\frac{ac}{5b}$

12. $\frac{1}{4}$

13. $\frac{2}{15}$

14. 6

15. $\frac{3}{10}$

16. $\frac{7}{16}$

17. 1

18. $\frac{5}{2}$

19. $\frac{xy}{3z}$

20. Answers will vary.

Addition and Subtraction of Fractions

21. $\frac{12}{11}$

22. $\frac{2}{3}$

23. 12

24. 60

25. $\frac{8}{12}, \frac{3}{12}$

26. $\frac{16}{60}, \frac{9}{60}$

27. $\frac{11}{12}$

28. $\frac{5}{12}$

29. $\frac{41}{42}$

An Application

30. $6\frac{7}{8}$ inches

1.3 Exponents and Order of Operations

Key Terms

1. exponential expression
2. base; exponent
3. parentheses, absolute values
   exponential expressions
   multiplication; division
   addition; subtraction

Natural Number Exponents

1. 6^5
2. $\left(\frac{1}{3}\right)^4$
3. a^7
4. 64
5. 100,000
6. $\frac{8}{27}$
7. 10^3
8. 2^5
9. 3^4

Order of Operations

10. 3
11. 8
12. 7
13. $\frac{1}{5}$
14. 6
15. 11
16. 1
17. 28

Translating Words to Expressions

18. $3^3 - 8 = 19$
19. $30 + 4 \cdot 2 = 38$
20. $\frac{4^3}{2^2} = 16$
21. $\frac{40}{10-2} = 5$

1.4 Real Numbers and the Number Line

Key Terms

1. irrational number
2. principal square root
3. origin
4. average
5. rational number
6. opposite; additive inverse
7. square root
8. absolute value
9. integers
10. real number

Signed Numbers

1. -11
2. $\frac{3}{8}$
3. -5
4. $-\frac{1}{5}$

Integers and Rational Numbers

5. natural number,
whole number,
integer,
rational number

6. integer,
rational number

7. whole number,
integer,
rational number

8. rational number

Square Roots

9. 8

10. 20

11. 2.449

Real and Irrational Numbers

12. irrational number

13. integer,
rational number

14. rational number

15. natural number,
whole number,
integer,
rational number

16. Average height: 69 inches
natural number, rational number

The Number Line

17.

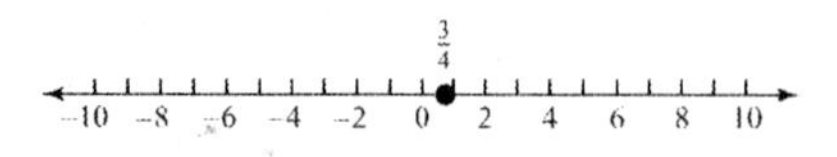

18.

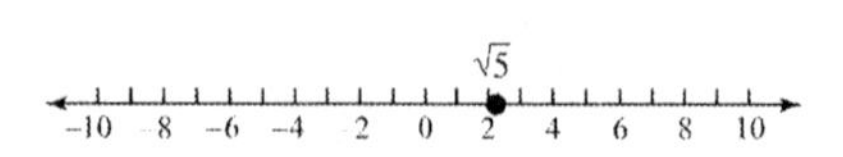

19.

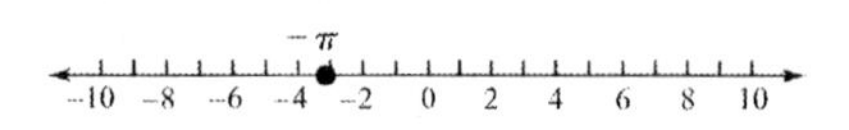

Absolute Value

20. 6.7

21. 4

22. 7

Inequality

23. $-3, -\sqrt{3}, 0, 2.2, \pi$

1.5 Addition and Subtraction of Real Numbers

Key Terms

1. less than or equal to; $a \le b$

2. difference

3. greater than; $a > b$

4. approximately equal

5. greater than or equal to; $a \ge b$

6. addends; sum

7. less than; $a < b$

Addition of Real Numbers

1. –32; 0

2. $\sqrt{3}$; 0

3. $-\frac{5}{6}, 0$

4. −13

5. $\frac{1}{20}$

6. –4.3

7. 8

8. −3

9. −4

10. −10

Subtraction of Real Numbers

11. −20

12. −12

13. −2.4

14. $\frac{2}{15}$

15. 9

16. $\frac{1}{4}$

17. −6.4

Applications

18. 172°F

19. $265

1.6 Multiplication and Division of Real Numbers

Key Terms

1. negative

2. reciprocal; muliplicative inverse

3. positive

4. dividend; divisor; quotient

Multiplication of Real Numbers

1. −24

2. $\frac{3}{10}$

3. 4.8

4. −72

5. 25

6. −25

7. −64

8. −64

Division of Real Numbers

9. −45

10. $\frac{1}{9}$

11. $-\frac{3}{8}$

12. 0

13. $0.8\overline{3}$

14. 0.4375

15. 1.125

16. $\frac{2}{25}$

17. $\frac{11}{40}$

18. $\frac{1}{200}$

19. $1.\overline{6}$; $1\frac{2}{3}$

20. 0.9375; $\frac{15}{16}$

Applications

21. **(a)** \$560,000
(b) \$320,000

22. 0.076

1.7 Properties of Real Numbers

Key Terms

1. identity property of 0

2. associative property for addition

3. multiplicative inverse property

4. distributive property

5. commutative property for multiplication

6. identity property of 1

7. associative property for multiplication

8. additive inverse property

9. commutative property for addition

Commutative Properties

1. $20+4$

2. $9 \cdot x$ or $9x$

Associative Properties

3. $2+(4+7)$

4. $(ab)c$

5. associative property for addition

6. commutative property for multiplication

7. commutative property for addition

Distributive Properties

8. $4a-20$

9. $-3b-24$

10. $-x+9$

11. $11-y$

12. $(7+4)x=11x$

13. $(2-10)a=-8a$

14. $(-6+3)y=-3y$

15. commutative and associative properties for addition

16. associative property for multiplication

17. distributive property

18. distributive property and commutative property for addition

Identity and Inverse Properties

19. identity property of 1

20. identity property of 0

21. multiplicative inverse property and identity property of 1

22. additive inverse property and identity property of 0

Mental Calculations

23. 1

24. 50

25. 195

26. 533

27. 5000 cubic feet

1.8 Simplifying and Writing Algebraic Expressions

Key Terms

1. additive identity
2. term
3. coefficient
4. like terms
5. multiplicative identity

Terms

1. Yes; -2
2. Yes; 4
3. No
4. Yes; 6

Combining Like Terms

5. unlike
6. like
7. unlike
8. like
9. $-\frac{7}{2}x$
10. $6y^2$
11. Cannot be combined

Simplifying Expressions

12. $-5+3x$
13. $7y-13$
14. a
15. $-3b-2$
16. $8x^2$
17. $-6t^3$
18. $4z-3$
19. $5a-5$

Writing Expressions

20. **(a)** $450w+600w+520w+700w = 2270w$

    (b) $108,960 \text{ ft}^2$

Chapter 2 Linear Equations and Inequalities

2.1 Introduction to Equations

Key Terms

1. addition property of equality
2. solution set
3. multiplication property of equality
4. equivalent
5. solution

The Addition Property of Equality

1. -6
2. 11
3. $\frac{7}{6}$
4. 1

The Multiplication Property of Equality

5. -9
6. -5
7. 9
8. $-\frac{1}{2}$
9. **(a)** $G = 24x$
   (b) 25 years

2.2 Linear Equations

Key Terms

1. linear equation
2. no solutions
3. infinitely many solutions
4. contradiction
5. one solution
6. identity

Basic Concepts

1. Yes; $a = 3, b = -2$
2. Not linear
3. Not linear
4. Yes; $a = \frac{2}{3}, b = -6$

Solving Linear Equations

5. -2

x	-3	-2	-1	0	1	2	3
$-2x-5$	1	-1	-3	-5	-7	-9	-11

6. $\frac{7}{4}$
7. -6
8. $\frac{5}{2}$
9. The year 2010

Applying the Distributive Property

10. $-\frac{15}{4}$

11. $\frac{13}{2}$

Clearing Fractions and Decimals

12. $-\frac{14}{15}$

13. $\frac{3}{4}$

14. 2.24

15. 0.4375

Equations with No Solutions or Infinitely Many Solutions

16. Infinitely many solutions

17. One solution

18. No solutions

2.3 Introduction to Problem Solving

Key Terms

1. percent change

2. fraction; decimal number

3. add, plus, more, sum, total, increase

4. subtract, minus, less, difference, fewer, decrease

5. multiply, times, twice, double, triple, product

6. divide, divided by, quotient, per

7. equals, is, gives, results in, is the same as

Steps for Solving a Problem

1. $4x+5=17;\ 3$

2. $\frac{1}{3}x+7=1;\ -18$

3. $15=2x-3;\ 9$

4. 17, 18, 19

5. 16,000

Percent Problems

6. $\frac{33}{100}$, 0.33

7. $\frac{1}{40}$, 0.025

8. $\frac{1}{125}$, 0.008

9. 13.7%

10. 12.5%

11. 161%

12. 20%

13. $55,775

14. 9.375 billion

Distance Problems

15. 570 mph

16. 8 mph

Other Types of Problems

17. 30 oz

18. $5000 at 9%; $1500 at 11%

2.4 Formulas

Key Terms

1. area
2. area
3. volume
4. circumference
5. perimeter
6. degree
7. area
8. circumference

Formulas from Geometry

1. **(a)** 123,500 square feet
   (b) approximately 2.8 acres
2. 36°; 72°; 72°
3. Circumference: $15\pi \approx 47.1$ centimeters
   Area: $56.25\pi \approx 177$ square centimeters
4. 10,275 square centimeters
5. Volume: 120 cubic meters
   Surface Area: 188 square meters
6. **(a)** approximately 19.63 cubic inches
   (b) approximately 10.9 fluid ounces

Solving for a Variable

7. **(a)** $l = \dfrac{P-2w}{2}$
   (b) $l = 8$ inches
8. $h = \dfrac{2A}{a+b}$
9. $x = \dfrac{yz}{z-y}$

Other Formulas

10. 2.58
11. **(a)** $C = \dfrac{5}{9}(F-32)$
    (b) 25°C

2.5 Linear Inequalities

Key Terms

1. interval notation
2. solution set
3. set-builder notation
4. linear inequality
5. solution

Solutions and Number Line Graphs

1. 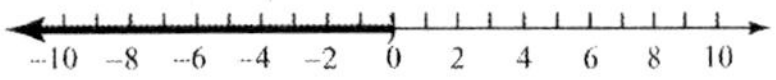 −10 −8 −6 −4 −2 0 2 4 6 8 10
2. 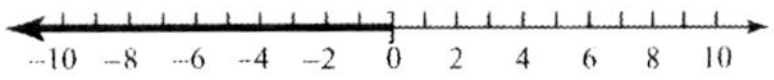 −10 −8 −6 −4 −2 0 2 4 6 8 10
3. 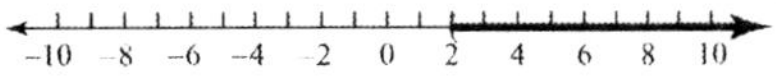 −10 −8 −6 −4 −2 0 2 4 6 8 10
4. 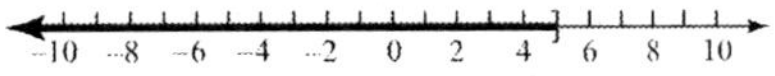 −10 −8 −6 −4 −2 0 2 4 6 8 10
5. $(-\infty, 7)$
6. $[-4, \infty)$
7. $(0, \infty)$
8. Yes
9. No
10. $x = 0$
11. $x < 0$
12. $x \le 0$
13. $x > 0$

The Addition Property of Inequalities

14. $x \le -9$

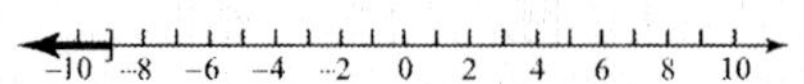

15. $x > 3$

16. $x \le 2$

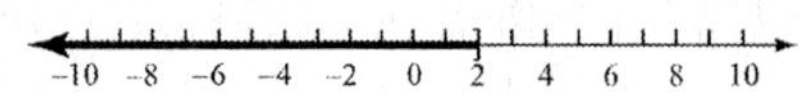

The Multiplication Property of Inequalities

17. $x \ge -7$

18. $x > 9$

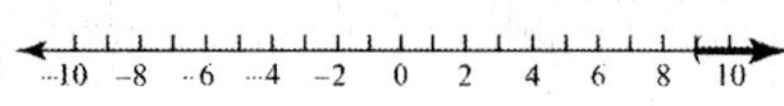

19. $\left\{x \middle| x < \frac{2}{3}\right\}$

20. $\{x | x \le 8\}$

21. $\{x | x > -3\}$

Applications

22. $x < -10$

23. $x \le 90$

24. $x \ge 3.5$

25. Altitudes greater than 4 miles

26. **(a)** $C = 4200 + 180x$
(b) $R = 240x$
(c) $P = 60x - 4200$
(d) 70 items

Chapter 3 Graphing Equations

3.1 Introduction to Graphing

Key Terms

1. ordered pair
2. line graph
3. x-coordinate; y-coordinate
4. origin
5. scatterplot
6. quadrants
7. rectangular coordinate system
8. x-axis; y-axis

The Rectangular Coordinate System

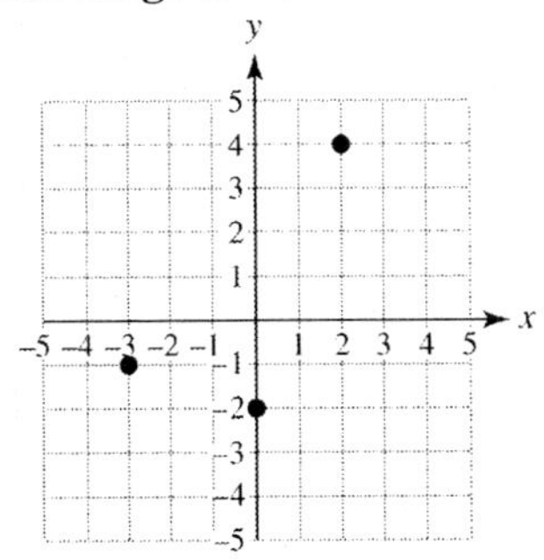

1. I
2. III
3. None; y-axis
4. 16 hours; 27 hours
5.

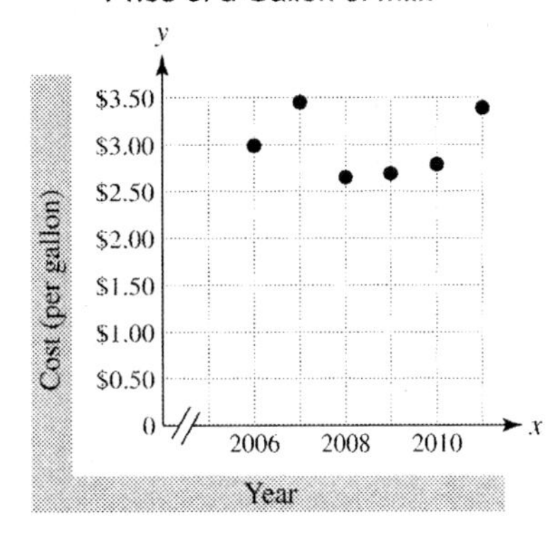

6.

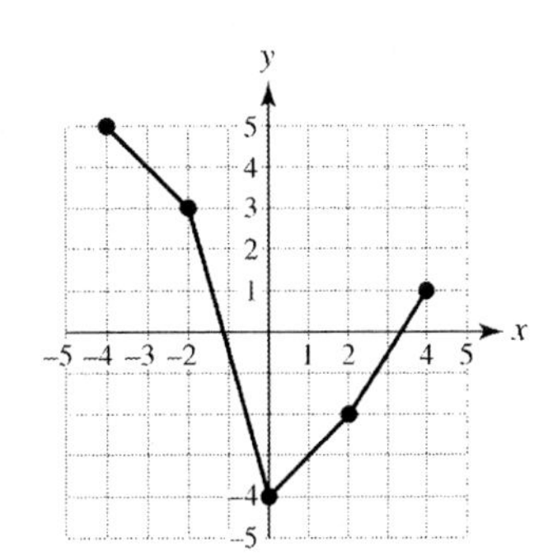

7. **(a)** no
   (a) 1960: about 115
   2000: about 150
   (a) approximately 30%

3.2 Linear Equations in Two Variables

Key Terms

1. linear equation in two variables
2. infinitely many
3. table
4. solution
5. standard form

Basic Concepts

1. No

2. No

3. Yes

Tables of Solutions

4.

x	−4	−2	0	2
y	−8	−2	4	10

5.

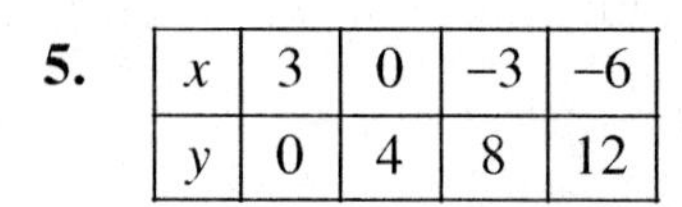

x	3	0	−3	−6
y	0	4	8	12

6. **(a)**

t	2	4	6	8	10
P	12.0	12.8	13.6	14.4	15.2

(b) 14.4 million

Graphing Linear Equations in Two Variables

7.

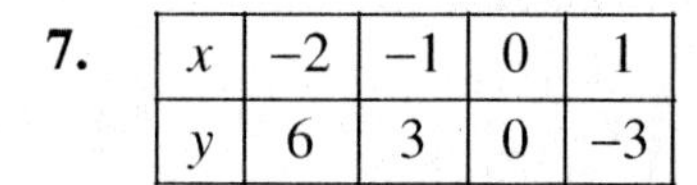

x	−2	−1	0	1
y	6	3	0	−3

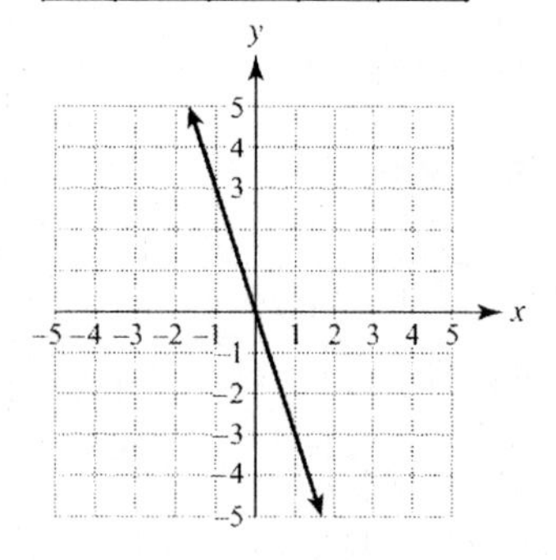

8. $y = -\dfrac{1}{2}x + 3$

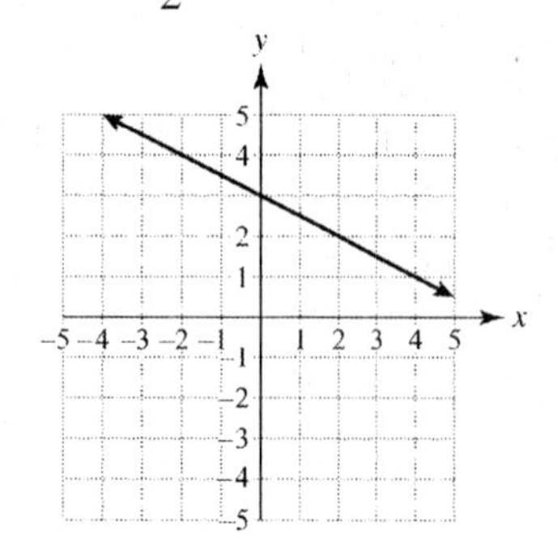

9. $x + y = -2$

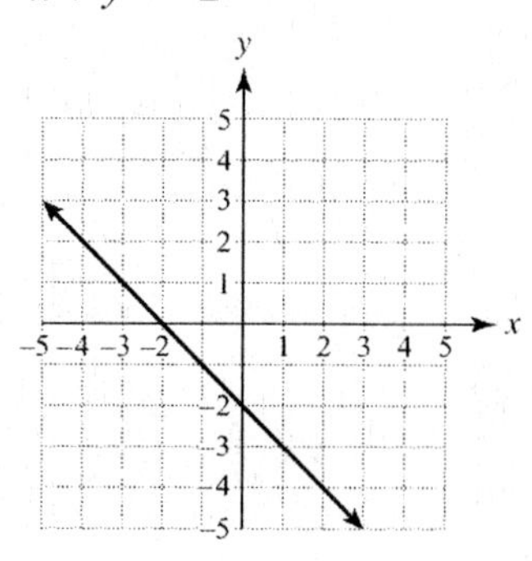

10. $y = \dfrac{3}{5}x - 3$

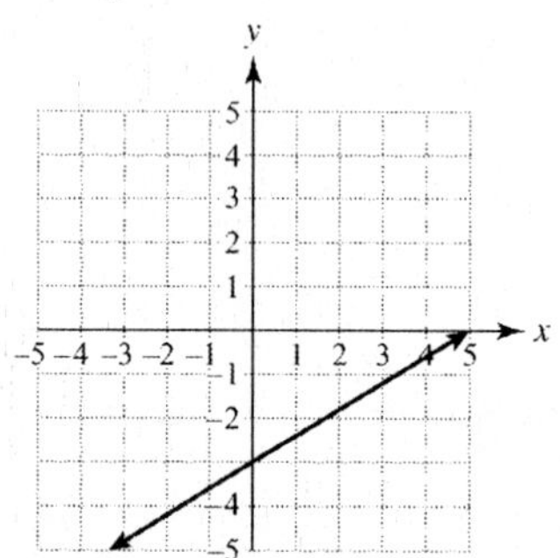

3.3 More Graphing of Lines

Key Terms

1. horizontal line

2. y-coordinate; y-axis; $x = 0$; y

3. vertical line

4. x-coordinate; x-axis; $y = 0$; x

Finding Intercepts

1. $-5x + 2y = 10$

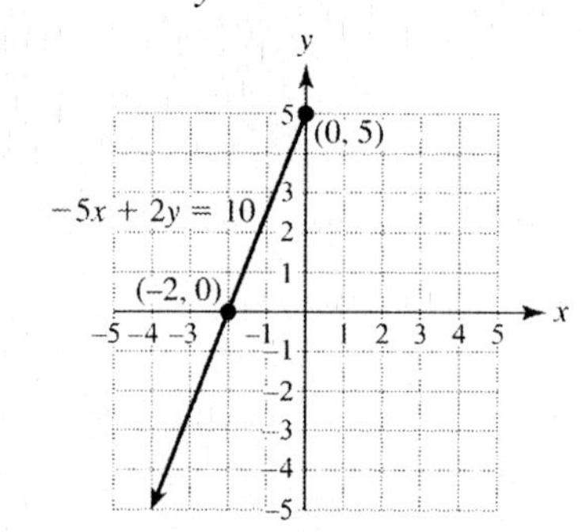

2.

x	-2	-1	0	1	2
y	0	1	2	3	4

x-intercept: -2
y-intercept: 2

3. **(a)**

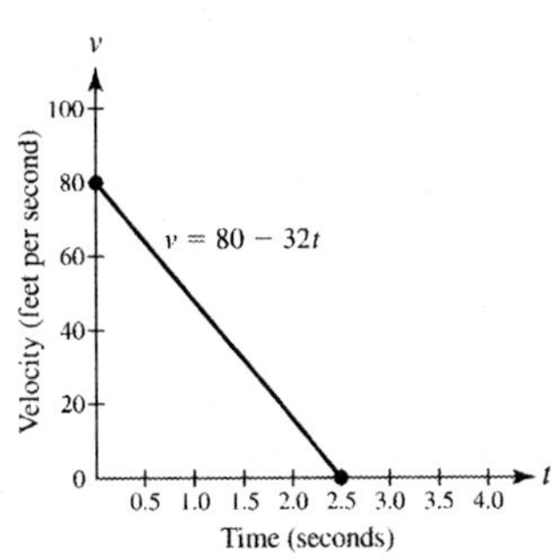

(b) The t-intercept indicates that the ball had a velocity of 0 feet per second after 2.5 seconds. The v-intercept indicates that the ball's initial velocity was 80 feet per second.

Horizontal Lines

4. y-intercept: -3

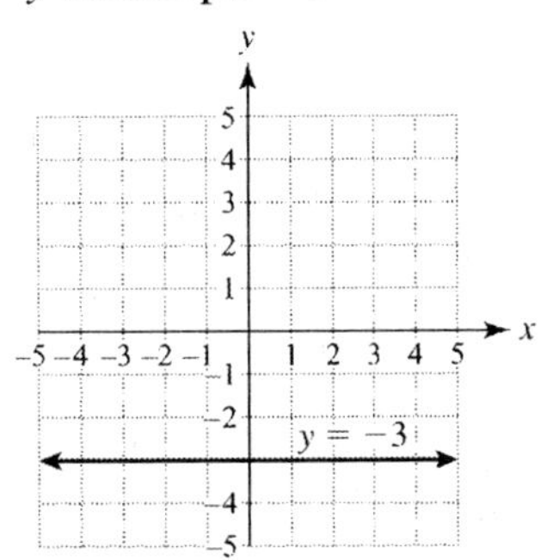

Vertical Lines

5. x-intercept: -1

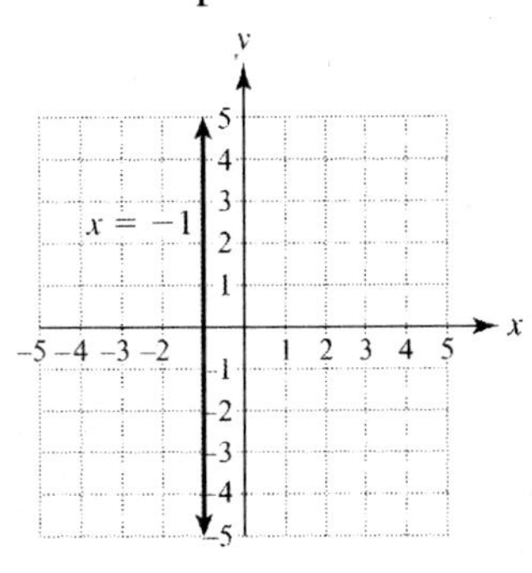

6. $y = 2$

7. $x = -3$

8. $x = -4$

9. $y = 1$

10. $x = 0$

3.4 Slope and Rates of Change

Key Terms

1. zero slope

2. negative slope

3. run

4. slope

5. rate of change

6. rise

7. positive slope

8. undefined slope

Finding Slopes of Lines

1. $m = -1$

2. $m = 0$

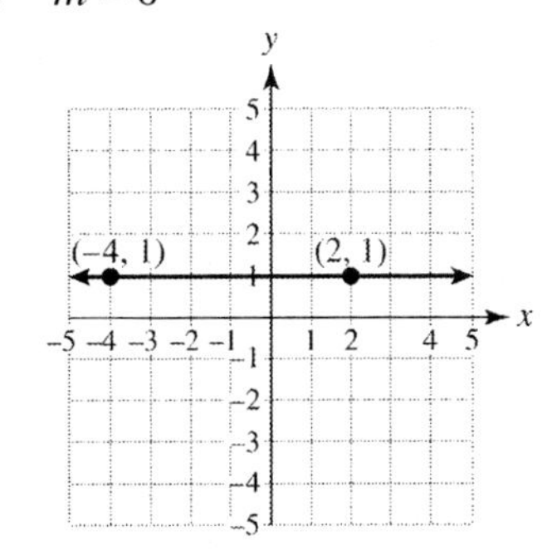

3. $m = 2$

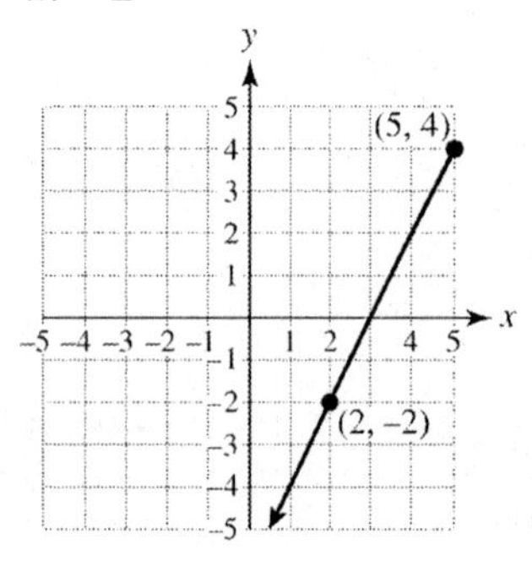

4. m is undefined

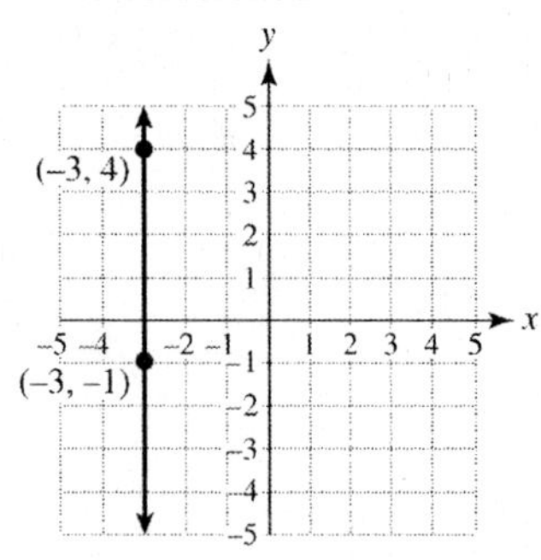

5. $m = 2$

6. m is undefined

7.

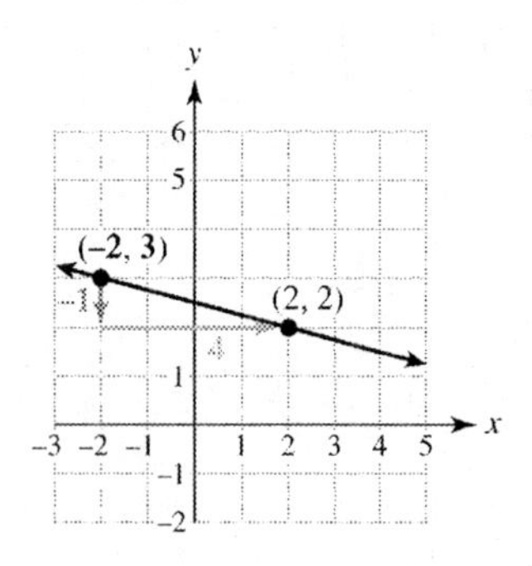

8.

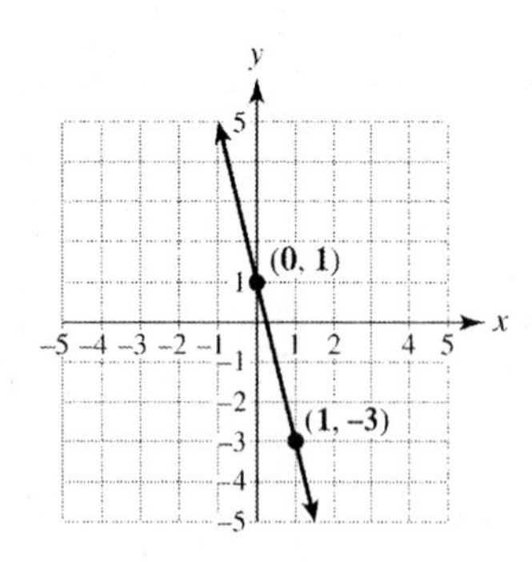

9.

x	–2	–1	0	1
y	4	1	–2	–5

Slope as a Rate of Change

10. **(a)** y-intercept: 80
The cyclist is initially 80 miles from home.
(b) After 2 hours the cyclist is 48 miles from home.
(c) $m = 16$; The cyclist is biking at 16 miles per hour.

11. **(a)** $m = 30$
(b) Profit increases, on average, by $30 for each additional game consolefter made.

12. **(a)**

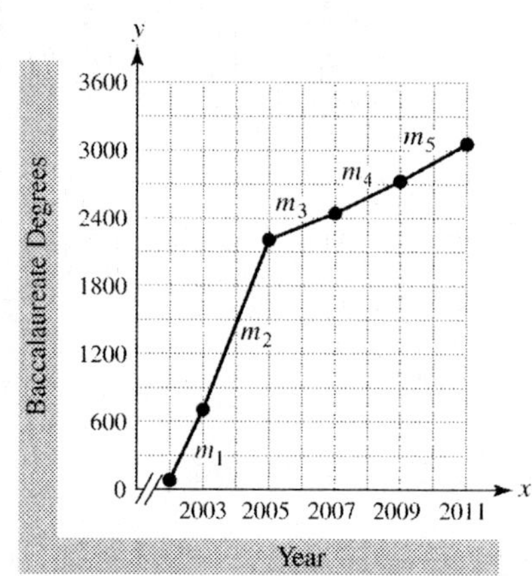

(b) $m_1 = 622$
$m_2 = 750$
$m_3 = 118.5$
$m_4 = 141.5$
$m_5 = 165.5$

(c) The slope indicates the increase in the number of Baccalaureate degrees awared.

13. (a)

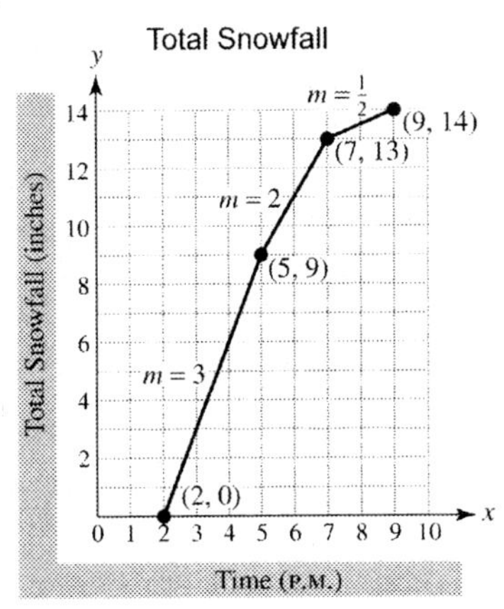

(b) The slope of each line segment represents the rate at which snow is falling.

3.5 Slope-Intercept Form

Key Terms

1. zero slope
2. parallel
3. slope
4. perpendicular
5. slope-intercept
6. negative reciprocals
7. undefined slope

Finding Slope-Intercept Form

1. $y = -2x + 2$
2. $y = x + 3$
3. $y = \frac{1}{3}x - 4$

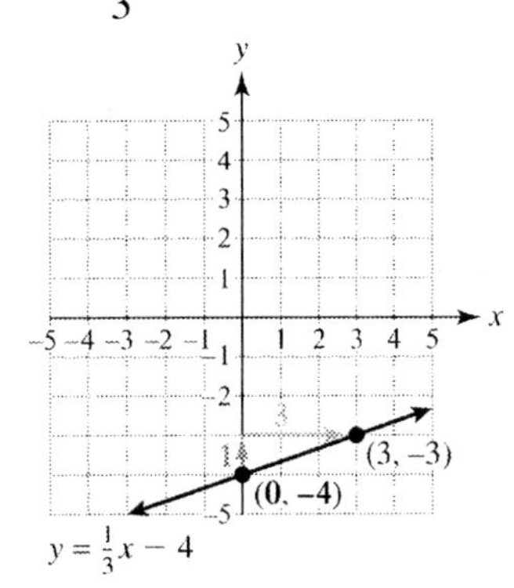

4. $y = \frac{5}{3}x + 5$; $m = \frac{5}{3}$; y-int: 5
5. $y = -\frac{1}{3}x + 2$; $m = -\frac{1}{3}$; y-int: 2
6. $y = 3x - 2$

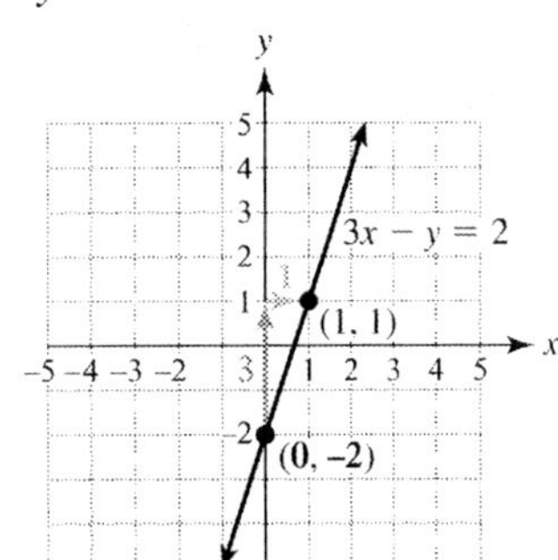

7. **(a)** \$274,000
   (b) $y = 120x + 34{,}000$
   (c) 3500

Parallel and Perpendicular Lines

8. $y = 2x + 5$

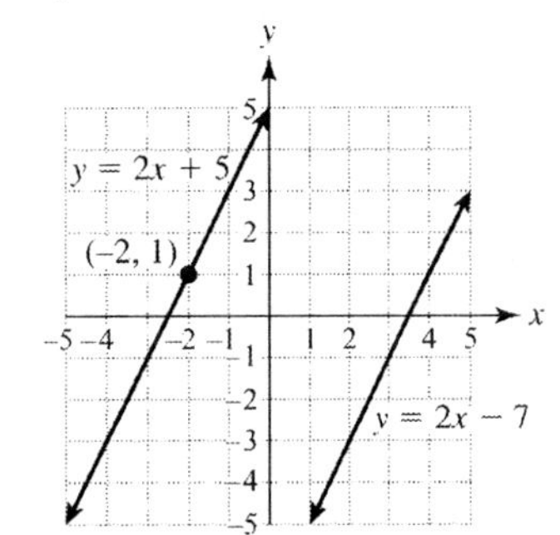

9. $y = \frac{1}{4}x$

10. $y = -\frac{3}{2}x$

11. $y = \frac{2}{5}x$

12. $y = -\frac{4}{3}x + 3$

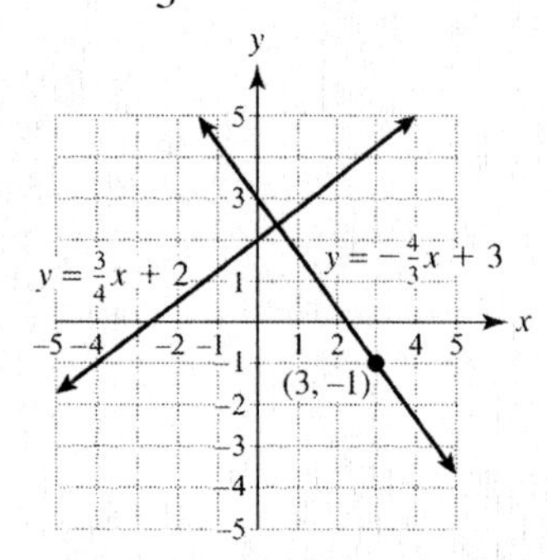

3.6 Point-Slope Form

Key Terms

1. point-slope

2. slope

3. slope-intercept

Derivation of Point-Slope Form

Writing Exercise: Answers will vary.

Finding Point-Slope Form

1. $y - 3 = -\frac{1}{2}(x + 2)$;
$y = -\frac{1}{2}x + 2$

2. $y - 0 = 3(x - 0)$;
$y = 3x$

3. $y + 2 = -\frac{1}{4}(x - 2)$; No

4. $y + 4 = -5(x - 2)$ or
$y - 1 = -5(x - 1)$

5. $y = -\frac{1}{2}x + 1$

6. $y = \frac{1}{4}x + 1$

7. $y = \frac{4}{3}x - 3$

8. $y = -3x + 1$

Applications

9. **(a)** $y - 56 = 3(x - 2006)$ or
$y - 74 = 3(x - 2010)$

(b) Slope $m = 3$ indicates that the salary increased, on average, by \$3000 per year.

(c) \$83,000

10. **(a)** 750 gallons per hour

(b) $y = -750x + 6000$

(c) The y-intercept is 6000 and indicates that the tank initially contained 6000 gallons of water. The x-intercept is 8 and indicates that the tank is empty after 8 hours.

(d)

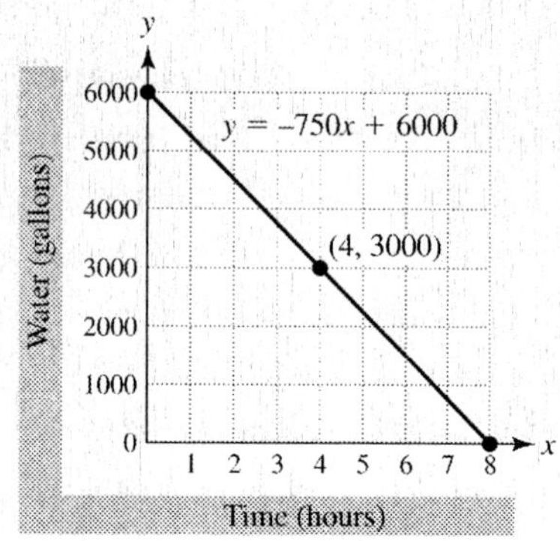

(e) After 4 hours, 3000 gallons of water remain in the tank.

3.7 Introduction to Modeling

Basic Concepts

1. No, at $x = 6$, $D = 610$.

Modeling Linear Data

2. **(a)**

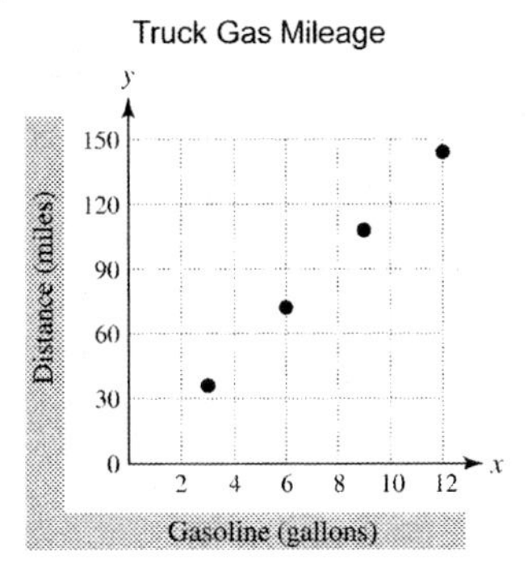

(b)

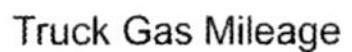

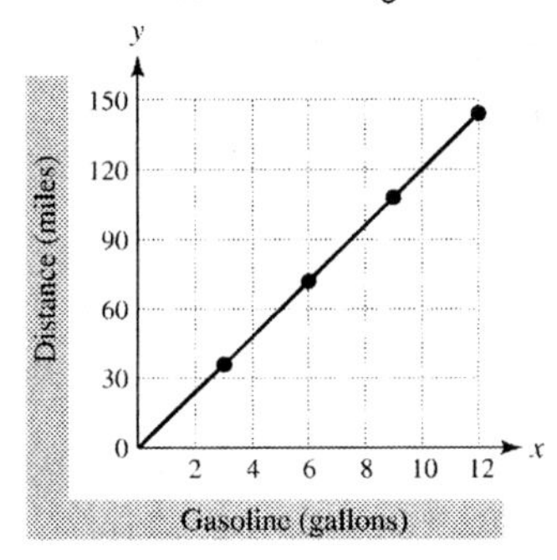

(c) $y = 12x$

The mileage of this truck is 12 miles per gallon.

(d) 192 miles

3. **(a)** No, the points are not collinear.

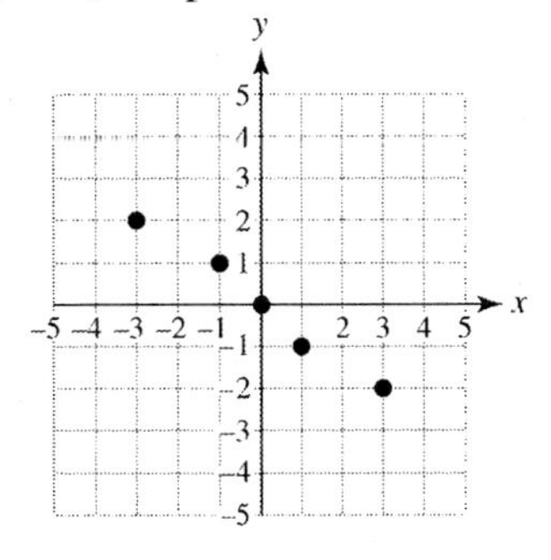

3. **(b)** $y = -\frac{2}{3}x$

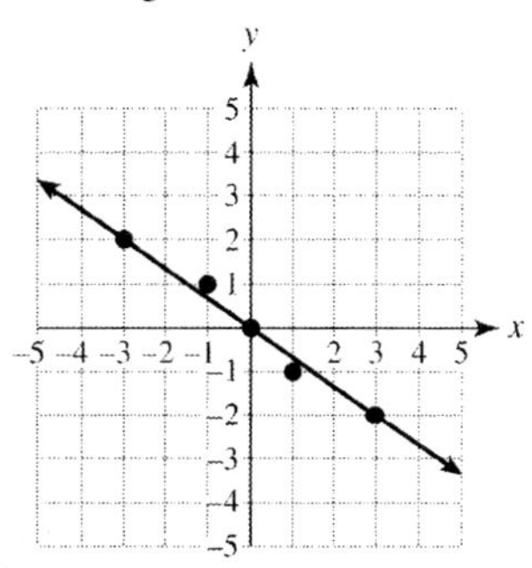

4. **(a)** $N = 1.9x + 18.8$

(b) Aproximately 26.4 million

5. $y = 2500$

6. $y = 28x + 700$

7. $y = -10x + 480$

Chapter 4 Systems of Linear Equations in Two Variables

4.1 Solving Systems of Linear Equations Graphically and Numerically

Key Terms

1. consistent; independent; intersecting
2. intersection-of-graphs
3. inconsistent; parallel
4. solution to a system
5. system of linear equations
6. consistent; dependent; identical

Basic Concepts

1. 5 hours

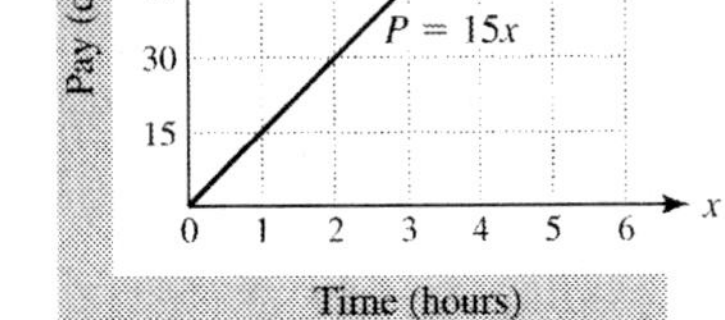

2. 2

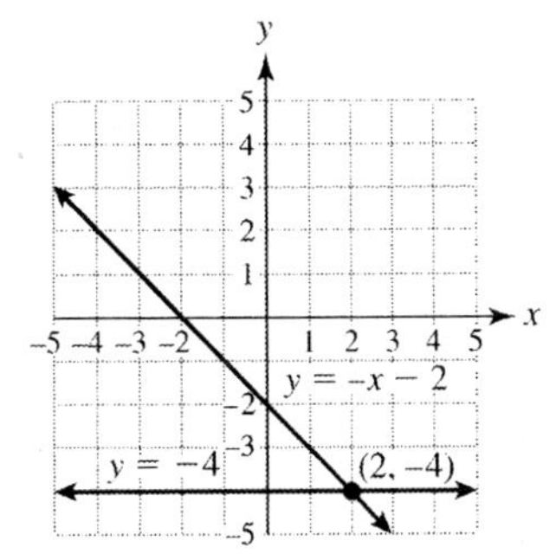

3. 0

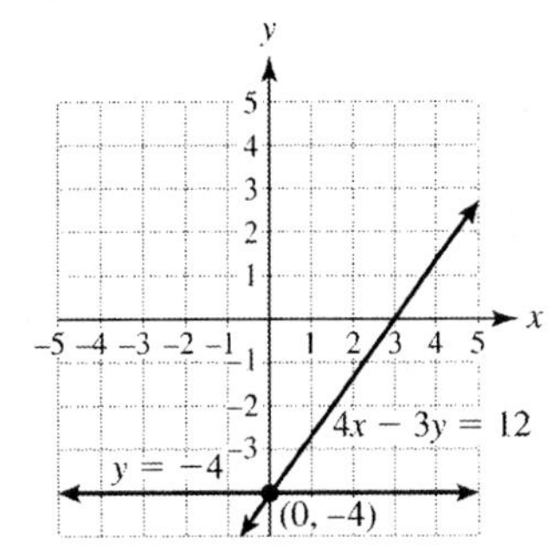

Solutions to Systems of Equations

4. infinitely many; consistent; dependent
5. 0; inconsistent
6. 1; consistent; independent
7. $(-5,-3)$
8. $(0,3)$

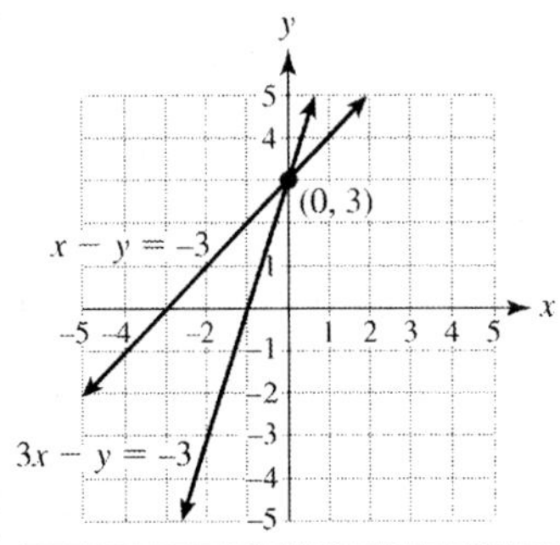

x	-3	-2	-1	0	1
$y=x+3$	0	1	2	3	4
$y=3x+3$	-6	-3	0	3	6

9. $(1,-2)$

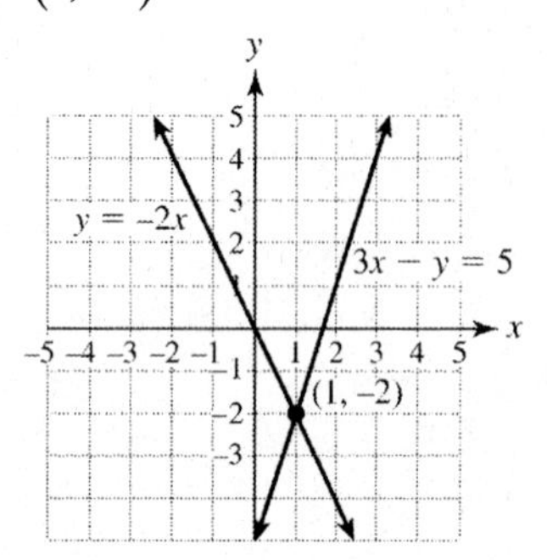

10. 5000 male students
7000 female students

4.2 Solving Systems of Linear Equations by Substitution

Key Terms

1. method of substitution

2. no solutions; parallel

3. infinitely many solutions; identical

The Method of Substitution

1. $(-2,-4)$

2. $\left(2,\frac{1}{2}\right)$

3. $(5,0)$

4. $(2,1)$

Recognizing Other Types of Systems

5. infinitely many solutions

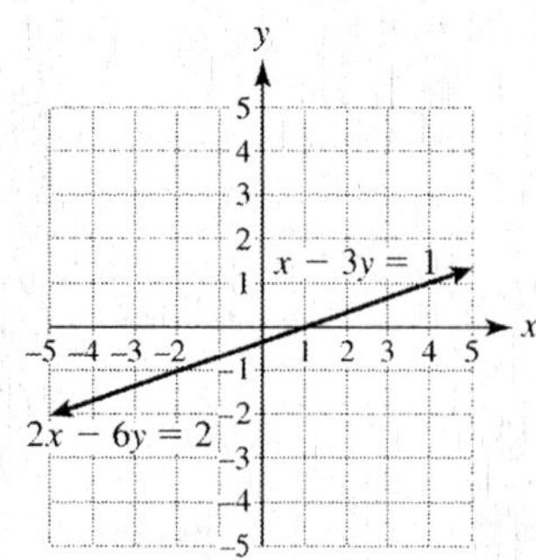

6. no solutions

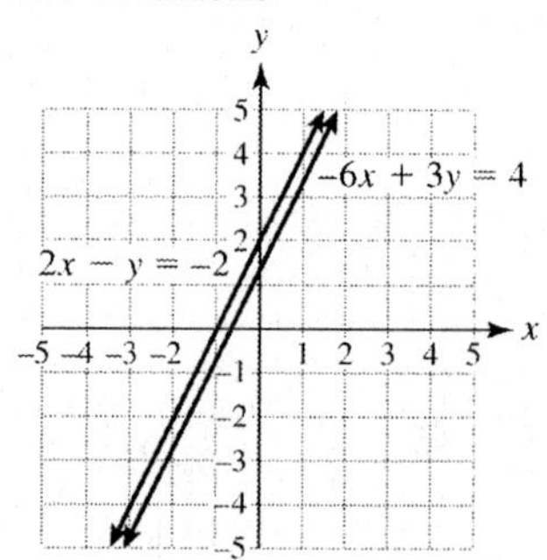

Applications

7. 300 radio ads
150 television ads

8. hors d'oeuvres: \$900
desserts: \$1600

9. speed of airplane: 450 mph
speed of wind: 30 mph

4.3 Solving Systems of Linear Equations by Elimination

Key Terms

1. no solutions; inconsistent

2. infinitely many solutions; consistent; dependent

3. elimination method

The Elimination Method

1. $(-2,1)$

2. $(2,2)$

3. $(-4,3)$

4. $\left(-\frac{1}{4},\frac{3}{4}\right)$

5. $(3,-5)$

6. $(1,-3)$

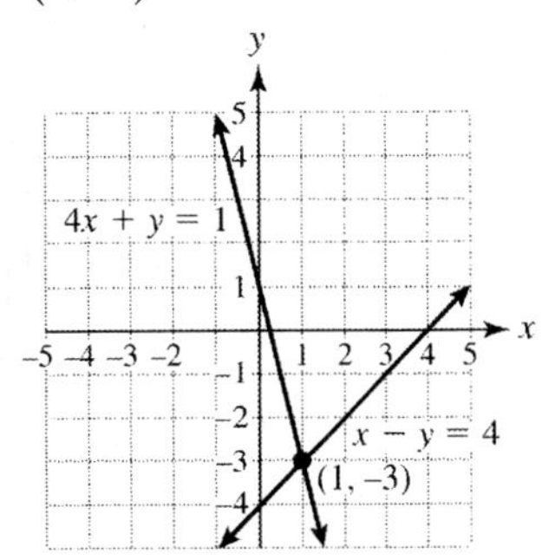

x	−2	−1	0	1	2
$y=x-4$	−6	−5	−4	−3	−2
$y=-4x+1$	9	5	1	−3	−7

Recognizing Other Types of Systems

7. no solutions

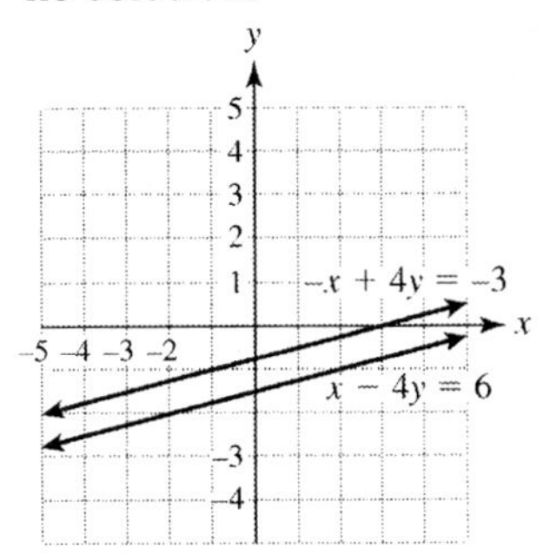

8. infinitely many solutions

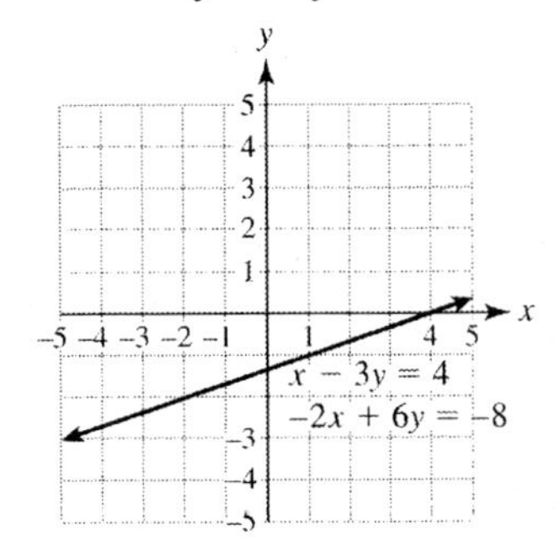

Applications

9. **(a)** 1100 women
(b) 700 men

10. 20 minutes on treadmill
10 minutes on elliptical machine

Understanding Concepts through Multiple Approaches

11. **(a)** $(-1,4)$

(b)

x	−2	−1	0	1	2
$y=-2x+2$	6	4	2	0	−2
$y=2x+6$	2	4	6	8	10

$(-1,4)$

(c) $(-1,4)$

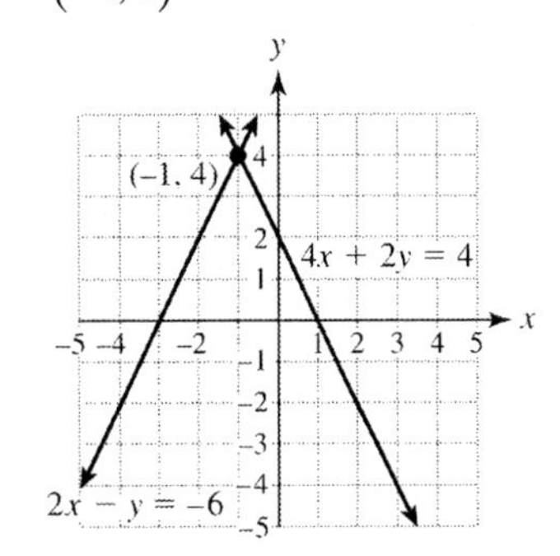

4.4 Systems of Linear Inequalities

Key Terms

1. test point

2. linear inequality

3. solution

4. system of linear inequalities

5. solution set

Solutions to One Inequality

1. $y \le 2$

2. $y > -2x$

3.

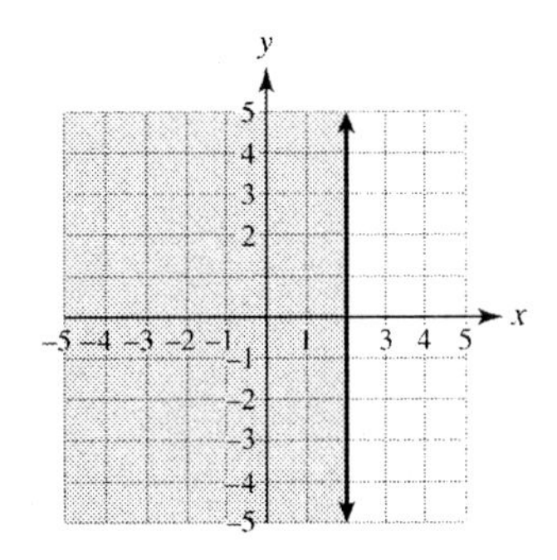

4.

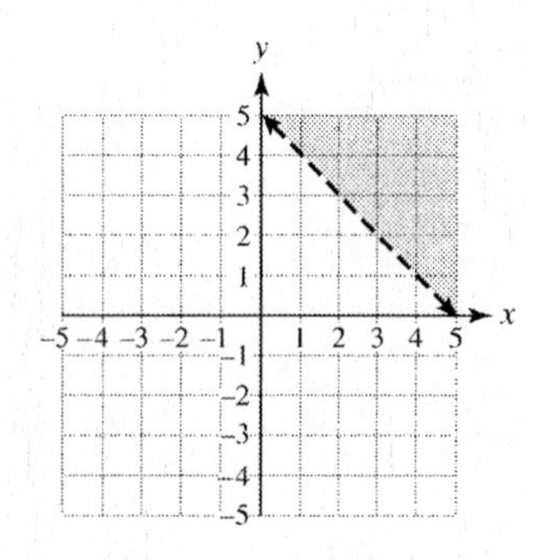

5.

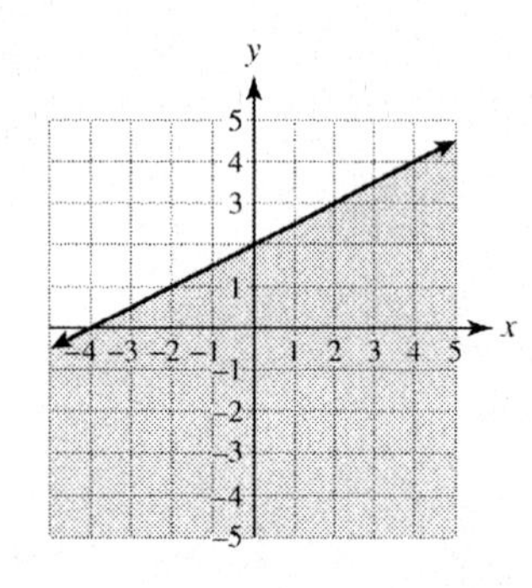

Solutions to Systems of Inequalities

6.

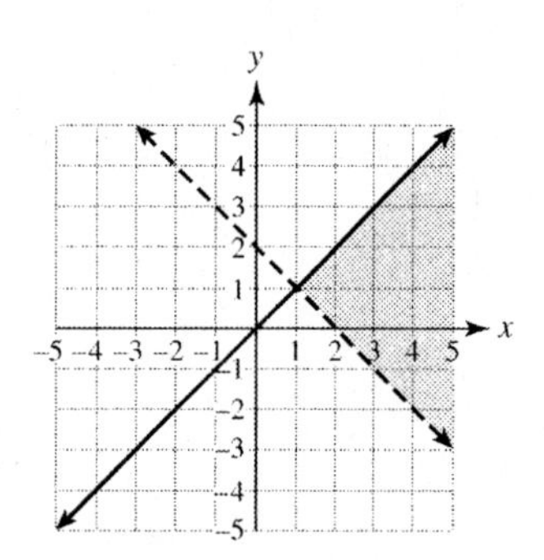

7.

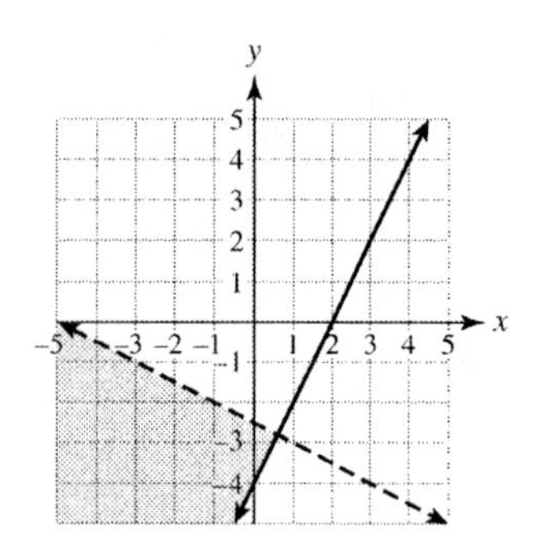

Applications

8.

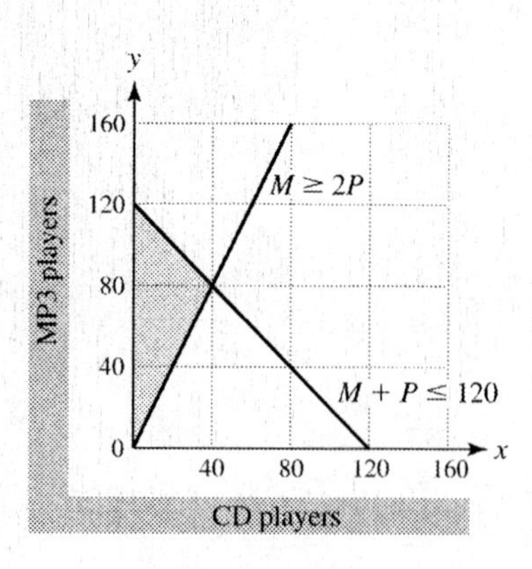

9. **(a)** The weight falls slightly outside of the recommended guidelines.
(b) 150 to 200 lb

Chapter 5 Polynomials and Exponents

5.1 Rules for Exponents

Key Terms

1. power to a power
2. quotient to a power
3. product
4. base; exponent
5. product to a power
6. zero exponent; undefined

Review of Bases and Exponents

1. 7
2. $\frac{1}{4}$
3. -81
4. 16

Zero Exponents

5. -1
6. 4
7. 1

The Product Rule

8. 243
9. x^8
10. $12x^7$
11. $3x^6+2x^5$
12. a^5
13. $(x+y)^4$

Power Rules

14. 2^{15}
15. x^{16}
16. $125t^3$
17. $-8b^9$
18. $3x^{12}y^8$
19. $-64z^{18}$
20. $\frac{27}{64}$
21. $\frac{x^6}{y^6}$
22. $\frac{36}{(a-b)^2}$
23. $200x^5$
24. $\frac{a^9b^3}{c^6}$
25. $-72x^{14}y^{12}$
26. **(a)** 3^2
    (b) $945,000

5.2 Addition and Subtraction of Polynomials

Key Terms

1. trinomial
2. degree
3. polynomial
4. like
5. monomial
6. polynomial
7. unlike
8. coefficient
9. binomial
10. degree

Monomials and Polynomials

1. Yes; 3; 1; 3
2. Yes; 4; 2; 5
3. No

Addition of Polynomials

4. Unlike
5. Like; $4x^3y$
6. Like; $\frac{7}{2}xy^2$
7. $-x+6$
8. $5y^2+3y+8$
9. $12x^3-10x^2+2x-1$
10. $4b^2-4b+13$

Subtraction of Polynomials

11. $3x-2$
12. $-x^2-10x+1$
13. $4x^3+5x^2-6x-4$
14. $3x^2+x-6$

Evaluating Polynomial Expressions

15. $4xy$; 320 in^3
16. 37 cents

5.3 Multiplication of Polynomials

Key Terms

1. coefficients
   variables
   product
2. product; like terms
3. binomials

Multiplying Monomials

1. $-12x^5$
2. $-20a^4b^6$

Review of the Distributive Properties

3. $6x-3$
4. $8x^2-10$
5. $-5x^2+2x$

Multiplying Monomials and Polynomials

6. $15x^3-20x$
7. $4x^3-x^2$
8. $-6x^2+8x-4$

9. $3x^6 - 6x^5 + 9x^3$

10. $20x^4y^3 + 4xy$

11. $-x^3y - xy^3$

Multiplying Polynomials

12. $x^2 + 8x + 15$

13. $2x^2 - 5x + 3$

14. $3 + 11x - 4x^2$

15. $3x^3 + x^2 - 2x$

16. $4x^3 - x^2 + x + 3$

17. $4a^3 + 4a^2b - 5ab^2 - 5b^3$

18. $b^6 - 4b^4 + 4b^2 - 1$

19. $x^3 + x^2 - 2x - 8$

20. **(a)** $h(h-2)(h+5) = h^3 + 3h^2 - 10h$
(b) 3900 cubic inches

5.4 Special Products

Key Terms

1. $a^2 - b^2$

2. $a^2 + 2ab + b^2$

3. $a^2 - 2ab + b^2$

Product of a Sum and Difference

1. $a^2 - 1$

2. $z^2 - 9$

3. $9r^2 - 25t^2$

4. $9m^4 - 4n^4$

5. 391

Squaring Binomials

6. $x^2 - 8x + 16$

7. $9x^2 - 12x + 14$

8. $1 + 12b + 36b^2$

9. $9a^4 - 6a^2b + b^2$

10. **(a)** $x^2 + 24x + 144$
(b) The area of the pool and sidewalk is 1024 ft^2.

Cubing Binomials

11. $125x^3 + 150x^2 + 60x + 8$

12. **(a)** $1 + 3r + 3r^2 + r^3$
(b) 1.125; The sum of money will increase by a factor of 1.125.

5.5 Integer Exponents and the Quotient Rule

Key Terms

1. $\frac{1}{a^n}$; reciprocal

2. a^{m-n}

3. **(a)** a^n
(b) $\frac{b^m}{a^n}$
(c) $\left(\frac{b}{a}\right)^n$

4. scientific notation

Negative Integers as Exponents

1. $\frac{1}{16}$

2. $\frac{1}{5}$

3. $\frac{1}{x^3}$

4. $\frac{1}{(a+b)^4}$

5. $\frac{1}{27}$

6. $\frac{1}{625}$

7. $\frac{1}{x^2}$

8. $\frac{1}{t^6}$

9. $\frac{1}{a^4b^4}$

10. $\frac{1}{x^3y^{11}}$

The Quotient Rule

11. $\frac{1}{9}$

12. $\frac{3}{a^4}$

13. $\frac{x}{y^5}$

Other Rules for Exponents

14. 81

15. $\frac{27}{16}$

16. $\frac{x^5}{3y^7}$

17. $\frac{25}{b^6}$

Scientific Notation

18. 417,000

19. 0.00023

20. 0.0582

21. 3.2×10^7

22. 2×10^{-5}

23. About 1.2×10^7 round trips

5.6 Division of Polynomials

Division by a Monomial

1. $x^4 - x$

2. $\frac{b^5}{2}-\frac{3b^2}{2}$

3. $3x-1+\frac{2}{x}$

4. $3a+4-\frac{2}{a}$

5. $l=x+3$

Division by a Polynomial

6. $3x-2+\frac{-2}{2x+3}$

7. $4x^2+4x+2+\frac{7}{x-1}$

8. $x-4+\frac{x+4}{x^2+4}$

Chapter 6 Factoring Polynomials and Solving Equations

6.1 Introduction to Factoring

Key Terms

1. factor
2. greatest common factor (GCF)
3. completely factored

Common Factors

1. $6(5x+3)$
2. $3x(3x-4)$
3. $4x(3x+2)$
4. $2y^2(5y-1)$
5. $4a(a^2-3a-2)$
6. $2x^2y(3x-y)$
7. $2a$; $2a(2a+3)$
8. $5x^2$; $5x^2(x^2-3)$
9. $3a^2b^2$; $3a^2b^2(2a-5b)$
10. $8t(5-2t)$

Factoring by Grouping

11. $(x-3)(4x-5)$
12. $(2t+3)(t^2+6)$
13. $(3x^2+4)(x-2)$
14. $(4+a)(x-y)$
15. $(5x^2-1)(x-3)$
16. $(2z^3-3)(z+6)$
17. $4(x^2+1)(2x-3)$
18. $x^2(2x^2-5)(2x+3)$

6.2 Factoring Trinomials I (x^2+bx+c)

Key Terms

1. standard form; leading coefficient
2. prime polynomial

Factoring Trinomials with Leading Coefficient 1

1. 4, 7
2. –8, 5
3. $(x+2)(x+5)$
4. $(x+3)(x+6)$
5. $(y+6)(y+7)$
6. $(b-3)(b-7)$
7. $(x-2)(x-6)$
8. $(y-5)(y+4)$
9. $(t+5)(t-8)$
10. $(x+6)(x-4)$

11. $(x-4)(x-3)$

12. Prime

13. $(x-7)(x+2)$

14. $5(x+2)(x+4)$

15. $2x^2(x+6)(x-1)$

16. Length: $x+5$;
Width: $x-2$

6.3 Factoring Trinomials II $\left(ax^2+bx+c\right)$

Factoring Trinomials by Grouping

1. $(2x+3)(x+5)$

2. $(4a+1)(a-3)$

3. $(5x-2)(2x-5)$

4. Prime

5. Prime

6. $5(y-4)(3y+1)$

7. $3x\left(x^2-6x-9\right)$

Factoring with FOIL in Reverse

8. $(3a+4)(a-5)$

9. $(2x+1)(x+5)$

10. $(x+8)(5x+1)$

11. $-(3x-1)(2x+5)$

12. $-(2x+7)(x-5)$

6.4 Special Types of Factoring

Key Terms

1. $(a+b)\left(a^2-ab+b^2\right)$

2. $(a-b)^2$

3. $(a-b)\left(a^2+ab+b^2\right)$

4. $(a+b)^2$

5. $(a-b)(a+b)$

6. $a^2+2ab+b^2$; $a^2-2ab+b^2$

Difference of Two Squares

1. $(x-4)(x+4)$

2. $(5x-2)(5x+2)$

3. $(9-5a)(9+5a)$

4. $(3x-10y)(3x+10y)$

Perfect Square Trinomials

5. $(x+6)^2$

6. $(3t+1)^2$

7. $(5x-4)^2$

8. $(x-7y)^2$

Sum and Difference of Two Cubes

9. $(x+5)\left(x^2-5x+25\right)$

10. $(a-2)\left(a^2+2a+4\right)$

11. $(x-6)\left(x^2+6x+36\right)$

12. $(4x-3)(16x^2+12x+9)$

13. $(4y+3)^2$

14. $(4b-5)(4b+5)$

15. $3x(2x-5)^2$

16. $a(3a-8b)(3a+8b)$

6.5 Summary of Factoring

Guidelines for Factoring Polynomials

greatest common factor (GCF)
grouping
a^2-b^2; difference of two squares
a^3-b^3; difference of two cubes
a^3+b^3; sum of two cubes
perfect square
$(a+b)^2$; perfect square trinomial
$(a-b)^2$; perfect square trinomial
grouping; FOIL
completely factored

Factoring Polynomials

1. $5x(x^2-4x+5)$

2. $4t^2(t-6)(t+6)$

3. $-5a(3a+1)^2$

4. $5(x-4)(x^2+4x+16)$

5. $2x^2(4x-1)(3x+2)$

6. $4(x-3)(x+3)(2x+1)$

7. $4ab(2a-3b)(2a+3b)$

8. $(3x^2+5)(4x+3)$

6.6 Solving Equations by Factoring I (Quadratics)

Key Terms

1. zero-product

2. quadratic polynomial

3. zeros

4. quadratic equation

5. standard form

The Zero-Product Property

1. –2, 0

2. 0

3. –1, 4

4. –5, 0, 3

Solving Quadratic Equations

5. –4, 0

6. –3, 3

7. 2, 3

8. $-\frac{3}{2}, \frac{4}{5}$

9. $-\frac{1}{2}, 5$

Applications

10. After 1.5 sec and 4 sec

11. **(a)** 177.8 ft
(b) About 23 mph
(c) About 23 mph

12. 50×70 pixels

Understanding Concepts through Multiple Approaches

13. **(a)** 36 mph

(b) 36 mph

6.7 Solving Equations by Factoring II (Higher Degree)

Polynomial with Common Factors

1. $-3(x-4)(x+1)$

2. $2x(x-5)(x-1)$

3. $-\frac{1}{3}, 0, \frac{1}{2}$

4. $-5, 0, 6$

5. 5 inches

Special Types of Polynomials

6. $(x-3)(x+3)(x^2+9)$

7. $(a^2+1)(a^2+5)$

8. $(r-t)^2(r+t)^2$

9. $(a-2b)(a+2b)(a^2+4b^2)$

10. $x(x-4)(x^2+4x+16)$

Chapter 7 Rational Expressions

7.1 Introduction to Rational Expressions

Key Terms

1. lowest terms
2. probability
3. undefined
4. vertical asymptote
5. rational expression; defined
6. basic principle

Basic Concepts

1. -1
2. $-\frac{1}{2}$
3. undefined
4. -1
5. 0
6. $-\frac{2}{3}$
7. $-3, 3$
8. None

Simplifying Rational Expressions

9. $-\frac{1}{3}$
10. $\frac{5}{8}$
11. $\frac{4}{t}$
12. $\frac{3}{4}$
13. $\frac{x-2}{x-1}$
14. $\frac{x+5}{2x-1}$
15. $-\frac{1}{3}$
16. -1
17. 1

Applications

18. (a)

x (cars/minute)	5	6	7	7.5	7.9	7.99
T (minutes)	$\frac{1}{3}$	$\frac{1}{2}$	1	2	10	100

(b) As the average traffic rate increases, the time needed to pass through the construction zone increases.

19. (a)

x (months)	0	6	12	36	72
P (thousands)	0.4	1.818	2.235	2.683	2.831

(b) 400 fish

(c) The fish population increased quickly at first but then leveled off.

20. (a) $\frac{4}{n}$

(b) $\frac{4}{100}=\frac{1}{25}$; $\frac{4}{1000}=\frac{1}{250}$; $\frac{4}{10{,}000}=\frac{1}{2500}$

(c) As the number of balls increases, the probability of picking the winning ball decreases.

7.2 Multiplication and Division of Rational Expressions

Key Terms

1. numerators; denominators
2. basic principle of fractions
3. reciprocal
4. lowest terms

Review of Multiplication and Division of Fractions

1. $\frac{6}{35}$
2. $\frac{5}{2}$
3. $\frac{1}{24}$
4. $\frac{4}{15}$
5. $\frac{1}{21}$
6. $\frac{3}{8}$

Multiplication of Rational Expressions

7. $\frac{4(2x+3)}{x(x+1)}$
8. $\frac{3x}{x-5}$
9. $x+3$
10. $\frac{x+4}{2(x-2)}$
11. **(a)** $D=\frac{120}{x}$

    (b) When $x=0.2,\ D=600$ feet

    When $x=0.6,\ D=200$ feet

    The car requires 3 times more distance to stop on an icy road than on dry pavement.

Division of Rational Expressions

12. $\frac{x+2}{6x}$
13. $\frac{x-2}{x^2+1}$
14. $\frac{x+3}{x+1}$

7.3 Addition and Subtraction with Like Denominators

Key Terms

1. like terms
2. subtract; numerators; denominator
3. greatest common factor
4. add; numerators; denominator

Review of Addition and Subtraction of Fractions

1. $\frac{5}{7}$
2. $\frac{3}{4}$
3. 1
4. $\frac{2}{3}$

Rational Expressions with Like Denominators

5. $\frac{9}{t}$

6. 1

7. $\frac{1}{a+5}$

8. $x+2$

9. $\frac{7}{ab}$

10. $\frac{1}{x-y}$

11. $\frac{4}{a-b}$

12. 1

13. $-\frac{2y}{2y-5}$

14. $\frac{1}{3x-4}$

15. $\frac{3x}{3x+4}$

16. $-\frac{1}{2}$

17. **(a)** $\frac{13}{n}$
(b) There are 13 chances in n that a defective calculator is chosen.

7.4 Addition and Subtraction with Unlike Denominators

Key Terms

1. prime factorization method
2. least common denominator
3. common multiple
4. listing method
5. least common multiple

Finding Least Common Multiples

1. $12a^2$
2. z^2+z
3. $(x-4)(x+5)$
4. $(x-1)(x-2)(x+3)$
5. $3x^2(x-5)(x+1)$

Review of Fractions with Unlike Denominators

6. $\frac{1}{36}$

7. $\frac{31}{35}$

Rational Expressions with Unlike Denominators

8. $\frac{15x}{9x^2}$

9. $\frac{2(x+y)}{x^2-y^2}=\frac{2x+2y}{x^2-y^2}$

10. $\dfrac{8y+5}{18y^2}$

11. $\dfrac{2x}{(x+2)(x-2)}$

12. $\dfrac{2x+3}{(x+3)^2}$

13. $\dfrac{-a^2+4a+8}{a(a+2)}$

14. $\dfrac{3x+1}{(x-1)(x+1)}$

15. $\dfrac{x^2-2x-1}{x(x-1)(x+1)}$

16. $\dfrac{3}{x}$

17. $\dfrac{4x-8}{(x-5)(x+1)}=\dfrac{4(x-2)}{(x-5)(x+1)}$

7.5 Complex Fractions

Key Terms

1. complex fraction
2. least common denominator
3. reciprocal
4. basic complex fraction

Simplifying Complex Fractions

1. $\dfrac{3}{4}$
2. $\dfrac{7}{10}$
3. $\dfrac{xy}{2}$
4. $3(z+2)$
5. $\dfrac{y-x}{y+x}$
6. $\dfrac{a^2+3}{a^2-3}$
7. $\dfrac{2x-1}{x-3}$
8. $\dfrac{ab}{b-a}$
9. $\dfrac{1}{x^2}$
10. $\dfrac{x}{5x+2}$
11. $\dfrac{3xy}{y+x}$

7.6 Rational Equations and Formulas

Key Terms

1. rational equation
2. algebraically
3. least common denominator; least common denominator; term; extraneous solution
4. visually
5. basic rational equation
6. numerically

Solving Rational Equations

1. $\frac{15}{4}$

2. $-\frac{2}{3}$

3. $-\frac{1}{2}, 4$

4. $\frac{7}{5}$

5. -10

6. $\frac{5}{2}$

7. No solutions

Rational Expressions and Equations

8. Expression; $\frac{x^2-2}{x+2}$; $\frac{7}{3}$

9. Equation; $x=\frac{1}{3}$

Graphical and Numerical Solutions

10. $-1, 3$

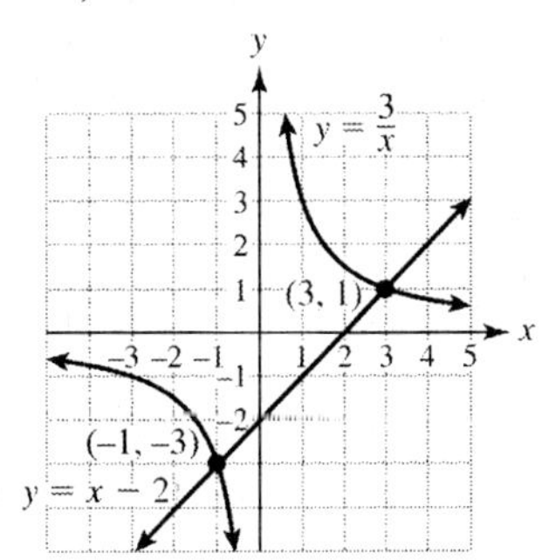

$-1, 3$

x	-3	-2	-1	0	1	2	3
$y_1 = \frac{3}{x}$	-1	$-\frac{3}{2}$	-3	$--$	3	$\frac{3}{2}$	1
$y_2 = x-2$	-5	-4	-3	-2	-1	0	1

Solving a Formula for a Variable

11. (a) 150 miles
    (b) $t=\frac{d}{r}$
    (c) 3 hours, 20 minutes

12. $h=\frac{2A}{b}$

13. $T=\frac{PV}{nR}$

14. $b_2=\frac{2A-b_1h}{h}$ or $b_2=\frac{2A}{h}-b_1$

Applications

15. $\frac{6}{5}$ hour, or 1 hr 12 min

16. $\frac{8}{5}$ hour, or 1 hr 36 min

Understanding Concepts through Multiple Approaches

17. (a) $x=4$
    (b)

x	-2	0	2	4	6
$y=\frac{2}{x-3}$	$-\frac{2}{5}$	$-\frac{2}{3}$	-2	2	$\frac{2}{3}$
$y=2$	2	2	2	2	2

(c)

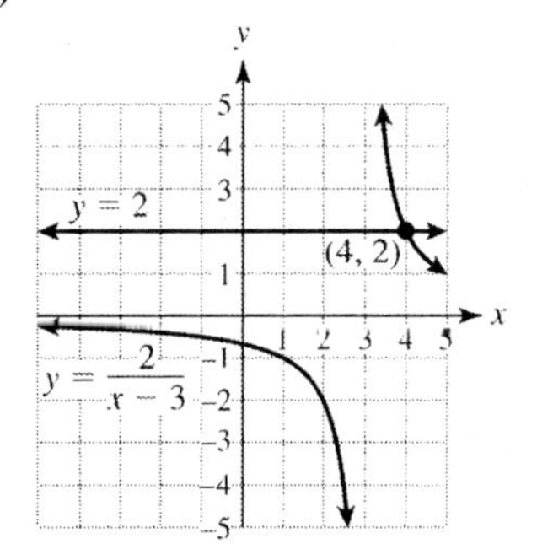

7.7 Proportions and Variation

Key Terms

1. inversely proportional; varies inversely
2. ratio
3. directly proportional; varies directly
4. proportion
5. varies jointly
6. constant of proportionality; constant of variation

Proportions

1. 1.75 in.
2. 16 feet

Direct Variation

3. $y = \frac{20}{3}$
4. $1652
5. **(a)** No
   (b) No

Inverse Variation

6. $y = 5$
7. **(a)** No
   (b) Yes
   (c) $y = 3$

Analyzing Data

8. Inverse
9. Neither
10. Direct
11. 120 lb

Chapter 8 Radical Expressions

8.1 Introduction to Radical Expressions

Key Terms

1. Pythagorean theorem
2. cube root
3. negative square root
4. perfect square
5. radical sign; radicand
6. square root
7. distance
8. principal square root

Square Roots

1. ±5
2. ±3.162
3. 8
4. 15
5. $\frac{2}{3}$
6. (a) $\frac{0.87}{\sqrt{0.6}} \approx 1.123$ step/sec

   (b) $\frac{0.87}{\sqrt{4}} = 0.435$ step/sec
7. real number, rational number
8. none
9. real number, irrational number

Cube Roots

10. 4
11. –6
12. 2.621

The Pythagorean Theorem

13. $c = 5$ inches
14. $a = 6$ inches
15. 64 feet

The Distance Formula

16. $\sqrt{32} \approx 5.66$
17. 37.5 miles

Graphing (Optional)

18.

x	$\sqrt{x-3}$
3	0
4	1
7	2
12	3

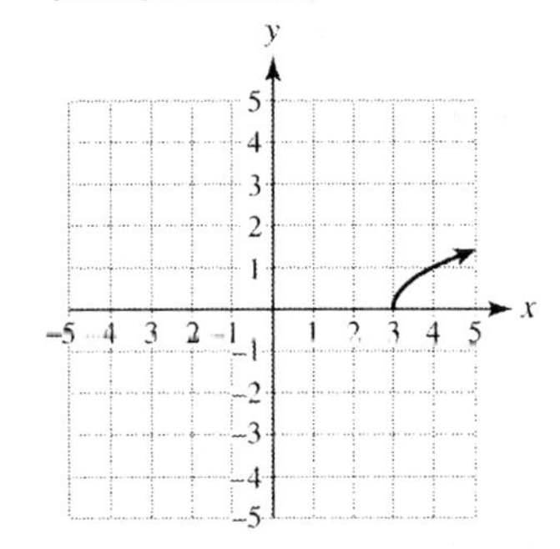

8.2 Multiplication and Division of Radical Expressions

The Product Rule

1. 6
2. 9
3. $\sqrt{14ab}$
4. $\sqrt{15xy}$
5. $\sqrt{2}\cdot\sqrt{7}$
6. $\sqrt{5}\cdot\sqrt{a}$
7. $\sqrt{3}\cdot\sqrt{11}$

Simplifying Square Roots

8. $2\sqrt{7}$
9. $5\sqrt{3}$
10. $11\sqrt{2}$
11. x^3
12. $a^2\sqrt{a}$
13. rt^2
14. $6z$
15. $7x\sqrt{x}$
16. $5x^3y^2\sqrt{y}$
17. $10a^2$

The Quotient Rule

18. $\frac{5}{7}$
19. $\frac{3}{z^3}$
20. $\frac{\sqrt{7}}{8x}$
21. $\sqrt{3}$
22. 5
23. $\frac{6}{ab}$

8.3 Addition and Subtraction of Radical Expressions

Addition of Radical Expressions

1. $4\sqrt{2}$, $5\sqrt{2}$
2. Not possible
3. $9\sqrt{3}$
4. $7\sqrt[3]{10}$
5. $26\sqrt{2}$
6. $8\sqrt{5}$
7. $6\sqrt{a}$
8. $10\sqrt[3]{rt}$
9. $17\sqrt{x}$
10. $11\sqrt{3}$

11. $6x\sqrt{x}+4x$

12. $10\sqrt{6}$ inches

13. **(a)** 850 cubic feet per second
(a) $2500\sqrt{m}$

Subtraction of Radical Expressions

14. $5\sqrt{2}$

15. $4\sqrt{5}$

16. $2\sqrt{x}$

17. $(9a-2)\sqrt{a}$

18. $6y\sqrt{x}$

19. $8\sqrt[3]{x}$

8.4 Simplifying Radical Expressions

Key Terms

1. like radicals
2. rationalizing the denominator
3. conjugate

Simplifying Products

1. 1
2. $x+3\sqrt{x}-28$

Rationalizing the Denominator

3. $\frac{2\sqrt{5}}{5}$

4. $\frac{\sqrt{7}}{3}$

5. $-\frac{2\sqrt{x}}{x}$

6. $\frac{\sqrt{5y}}{y}$

7. $2-\sqrt{3}$

8. $\sqrt{5}+3$

9. $\sqrt{a}+\sqrt{b}$

10. $7-4\sqrt{3}$

11. $\frac{a+2\sqrt{ab}+b}{a-b}$

8.5 Equations Involving Radical Expressions

Key Terms

1. radical equation
2. squaring property for solving equations
3. extraneous solution

Solving Radical Equations

1. Approximately 605 feet
2. 31
3. 3; -2 is an extraneous solution
4. –46
5. 16; 1 is an extraneous solution

Solving an Equation for a Variable

6. **(a)** 30 mph
(b) 900 feet

7. **(a)** About 1.92 sec
(b) About 5 feet

Understanding Concepts through Multiple Approaches

8. **(a)** 33

(b) 33

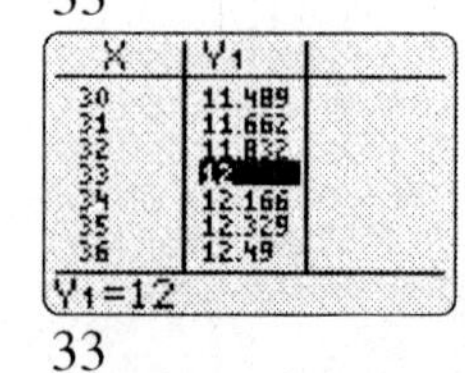

(c) 33

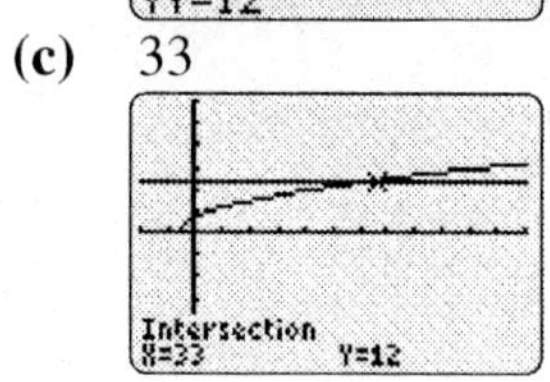

8.6 Higher Roots and Rational Exponents

Key Terms

1. nth root; odd root; even root

2. principal nth root; index

3. $\sqrt[n]{ab}$; $\dfrac{\sqrt[n]{a}}{\sqrt[n]{b}}$

4. $\sqrt[n]{a}$

5. $\sqrt[n]{a^m}$; $\left(\sqrt[n]{a}\right)^m$

6. $\dfrac{1}{a^{m/n}}$

Higher Roots

1. -4

2. 2

3. Not a real number

4. -3

5. -5

6. 3

7. -4

8. 1

9. $\dfrac{2}{3}$

10. $\dfrac{\sqrt[4]{8}}{5}$

11. -3

12. 2

Rational Exponents

13. $\sqrt{9} = 3$

14. $\sqrt[4]{81} = 3$

15. $\sqrt[3]{125} = 5$

16. $-\sqrt[4]{81^3} = -\left(\sqrt[4]{81}\right)^3 = -27$

17. $\sqrt[3]{(-64)^2} = \left(\sqrt[3]{-64}\right)^2 = 16$

18. $\sqrt[5]{32^3} = \left(\sqrt[5]{32}\right)^3 = 8$

19. **(a)** $L = 0.91\sqrt[3]{W}$

(b) About 1.04 m

20. $\dfrac{1}{\sqrt[5]{243^3}} = \dfrac{1}{\left(\sqrt[5]{243}\right)^3} = \dfrac{1}{27}$

21. $-\dfrac{1}{\sqrt[4]{16^3}} = -\dfrac{1}{\left(\sqrt[4]{16}\right)^3} = -\dfrac{1}{8}$

22. $\dfrac{1}{\sqrt[3]{(-125)^2}} = \dfrac{1}{\left(\sqrt[3]{-125}\right)^2} = \dfrac{1}{25}$

9.1 Parabolas

Key Terms

1. parabola; upward; downward
2. vertex; lowest; highest
3. axis of symmetry
4. x-coordinate
5. vertex; axis of symmetry
   upward; downward
   wider
   narrower

Graphing Parabolas

1. upward
2. downward
3. downward
4. $(3,4)$; $x=3$
5.

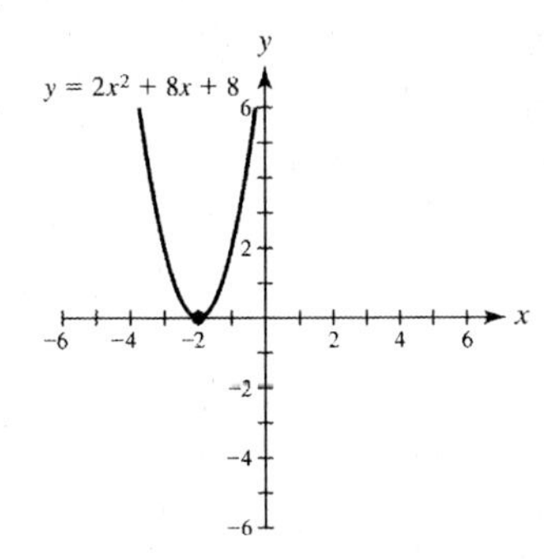

6. $(-1,4)$; $x=-1$

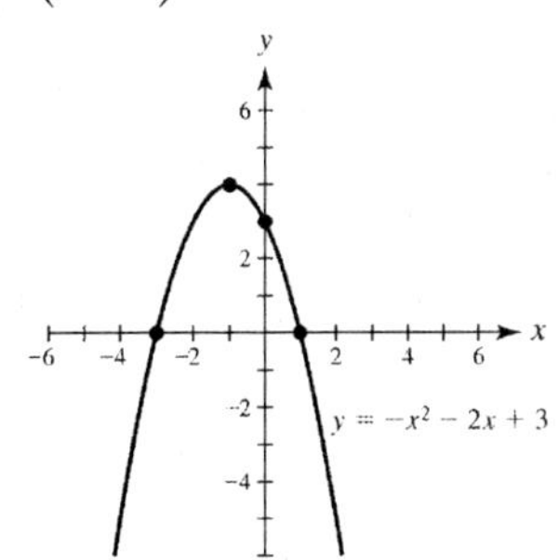

7. $(2,4)$; $x=2$

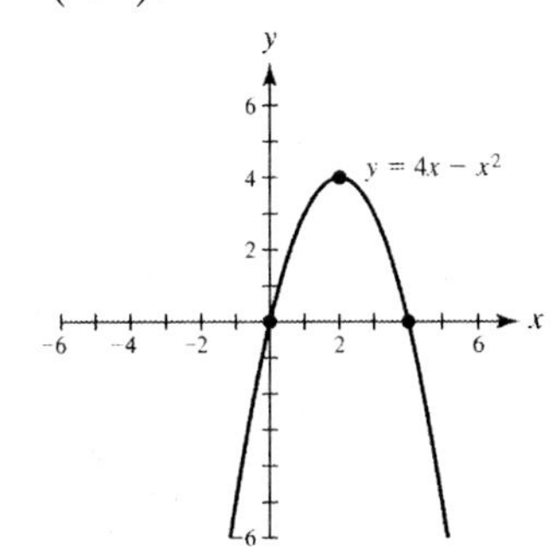

8. $(3,0)$; $x=3$

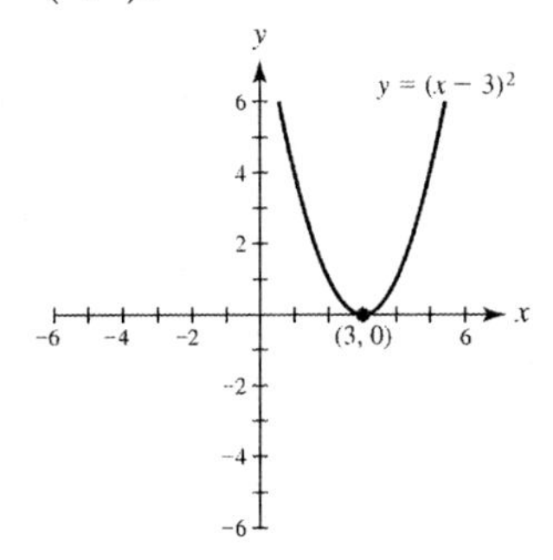

9. **(a)** $\left(\frac{7}{2}, 196\right)$
   (b) 3.5 seconds after the golf ball is hit, it reaches a maximum height of 196 feet.

The Graph of $y = ax^2$

10. The graph of $y = \frac{1}{2}x^2$ is wider than the graph of $y = x^2$.

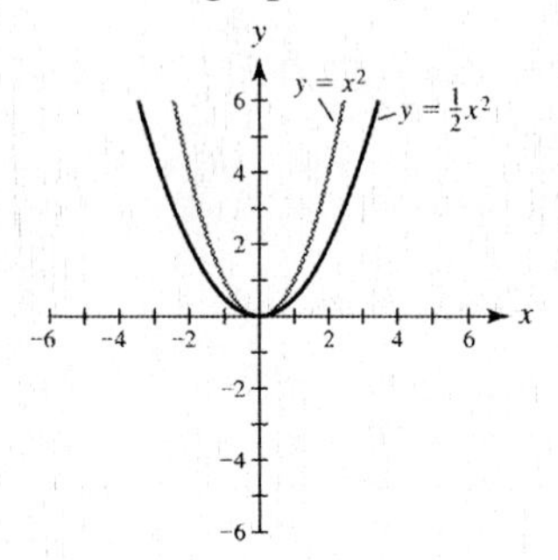

9.2 Introduction to Quadratic Equations

Key Terms

1. quadratic equation

2. square root property

Symbolic, Graphical and Numerical Solutions

1. No real solutions

2. $-2, -3$

3. $-1, 4$

The Square Root Property

4. $\pm\sqrt{5}$

5. $\pm\frac{4}{5}$

6. $-3, 9$

7. About 2.12 sec

8. $\approx \pm 4$

9.3 Solving by Completing the Square

Perfect Square Trinomials

1. Yes

2. No

3. Yes

Completing the Square

4. 49; $(x+7)^2$

5. $4 \pm \sqrt{14}$

6. $\frac{-7 \pm \sqrt{69}}{2}$

7. $\frac{-1 \pm \sqrt{13}}{3}$

9.4 The Quadratic Formula

Key Terms

1. quadratic formula

2. discriminant

3. two; one; no

Solving Quadratic Equations

1. $a = 2,\ b = 5,\ c = -1$

2. $a = -3,\ b = -5,\ c = 4$

3. $a = 1,\ b = 0,\ c = -16$

4. $-\frac{3}{2}, 5$

5. $\frac{-1 \pm \sqrt{13}}{6}$

6. 3

7. No real solutions

8. $\dfrac{1 \pm \sqrt{33}}{4}$

The Discriminant

9. Two

10. (a) $a < 0$
(b) $-1, 5$
(c) positive

9.5 Complex Solutions

Key Terms

1. imaginary unit

2. complex number; standard form; real part; imaginary part

3. imaginary number

Basic Concepts

1. $3i$

2. $i\sqrt{5}$

3. $2i\sqrt{3}$

Addition, Subtraction, and Multiplication

4. $-10 + 9i$

5. $7 + i$

6. $-3 + 11i$

7. $-7 + 24i$

Quadratic Equations with Complex Solutions

8. $\pm 2i\sqrt{2}$

9. $\dfrac{3}{2} \pm i\dfrac{\sqrt{11}}{2}$

10. $\dfrac{3}{2} \pm i\dfrac{\sqrt{15}}{2}$

9.6 Introduction to Functions

Key Terms

1. function

2. domain; range

3. vertical line test; function

4. function notation; input; output

Representing a Function

1. -15

2. -4

3. 4

4. $f(1) = -3$

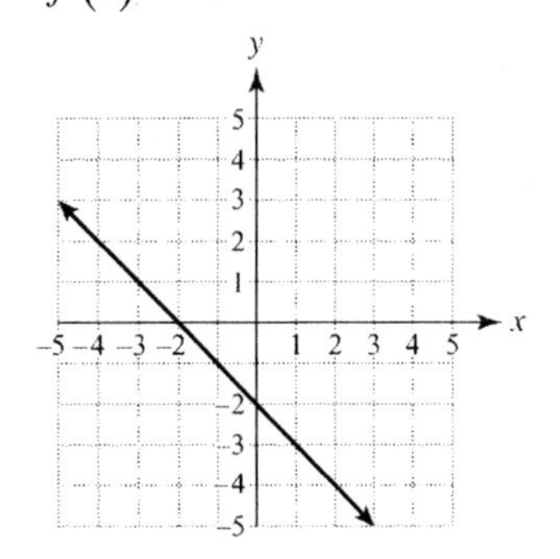

5. (a) $f(x) = x^2 - 4$

(b)

x	-2	-1	0	1	2
$f(x)$	0	-3	-4	-3	0

(c)

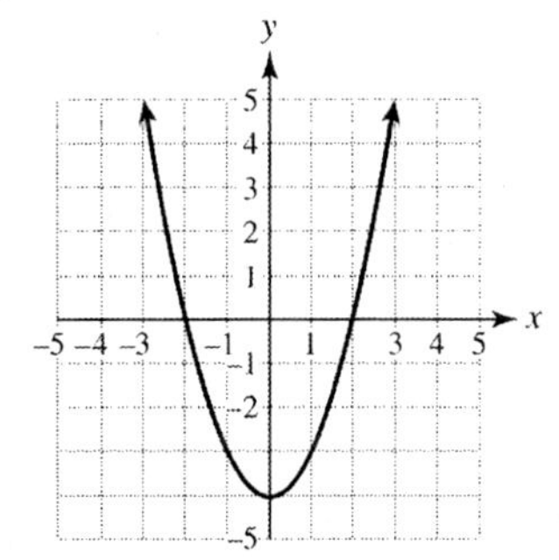

Definition of a Function

6. D: All real numbers

7. D: $x \neq 0$

8. D: $x \geq 4$

Identifying a Function

9. Yes

10. Yes

11. No